Falk's Claim

The
Life And Death
Of A Redwood Lumber Town

By Jon Humboldt Gates

Other Works By The Author

Night Crossings

Soviet Passage

Timezone/Lost Nations (CD)

FIRST PRINTING 1983
SECOND PRINTING 1985
THIRD PRINTING 1991
FOURTH PRINTING 1996

Cover photograph by Sy Beattie; the face of Noah Falk was superimposed by Peter Palmquist.

Published By:

moonstone
PUBLISHING
Post Office Box 911
Trinidad, California 95570

Printed in the United States
ISBN #1-878136-01-1

FOREWORD

The Redwood Region of Northern California had many pioneers. These included the Falk brothers—Noah, Elijah and Jonas. This story is especially of Noah, his sawmill, his town, and the people therein. All are heroes of a sort. It took heroic efforts to make a living in the "big tree country."

Never again will there be a place like Falk. Its main life spanned the years 1880-1930, and it was home to hundreds of people. Some stayed a few months, some most of a lifetime. Now there is little physical evidence of either mill or town, but it remains in the memory of many people.

Here, the reader shares some of the memories of real people who have told their stories of life and times at Falk.

Irving Wrigley
1982

Acknowledgments

I am deeply indebted to the dozens of people whose inspiration and assistance have proven invaluable in the completion of *Falk's Claim*. Foremost, I thank the people of Falk who have so willingly shared their time and past experience with me. Without their stories this book would not exist. Also, my highest respect and sincere appreciation goes to my mother and father for all they have given to my life, and to this project. Through my father, I have learned that storytelling can be a window into one's past; and my mother has demonstrated great patience and understanding while typing through reams of hand-scribbled manuscript drafts.

An immense gratitude is owed to Beverly Hanly and Teresa Henderson. These two women, besides taking many photographs, devoted much of themselves and their time toward helping me form a cohesive story from a knee-high pile of anecdotes and purple passages.

I thank Chad Hoopes for providing the original inspiration to uncover the town's history; Bob Mallett for a crash course on "How to use the English language," as well as assisting in the early development of the writing; Joey Mallett for her artwork and strong support in all matters of concern; Sayward Ayre for her photographs and proof reading; Sy Beattie, whose photographs speak without words; Tom Abate and Mia Ousley for their enthusiasm and confidence in bringing this work to public view; Sil, sister Susan, Robby, Airhead, and all of my long time friends who have encouraged me in my pursuits.

My final note of appreciation goes to the Wrigleys. I have known Ted and Irving for over 25 years, but only through this research did I learn that their family's 100 year history in the Elk River Valley had begun in the town of Falk. Ted and Irving, along with their sister, Ruth Braghetta, introduced me to most of the people I interviewed for this book.

Table of Contents

INTRODUCTION TO FALK'S CLAIM

MY FIRST VISIT TO FALK WAS IN THE SUMMER OF 1968. By then it was an old ghost town, dilapidated and practically hidden from view. But scores of unoccupied structures still stood within the dense growth of a returning forest. It was a humbling experience to walk along the main street of the town, through the mill and general store, the hotel-cookhouse, the post office, the gas station, and the many vacant houses. The area was strewn with the tattered remnants of past lives— abandoned vehicles, furniture, clothes, stoves, tools—all left behind but still in place. This mute energy captured my imagination.

During the afternoon of that first visit to the town, I found myself standing in the general store, surrounded by empty shelves and littered grocery receipts from the early 1900s. There was a stillness about the air, of the type which is found only in a well-preserved natural environment. This sensation led me to consider; since the beginning of history, people have struggled to survive against the elements, and at that moment, I was standing in a monument to past struggles. Except here, the forces of nature had won. I began to imagine the life that had moved through the aisles of the store, when suddenly a loud voice broke the silence. "What are you doing here!"

I turned around quickly and met the eyes of an old man who stood in the doorway, staring at me. "You've got no business here," he growled. "No one's allowed in the townsite." With a motion of his hardwood walking staff, he ushered me out of the building and away from the town. His gruff manner, however, was betrayed by a twinkling of friendliness in his eyes. This was Charlie Webb, the caretaker of the deserted community. I would later get to know him well, and his wife Loleta also. They provided me with my first information about Falk.

For several years I had heard stories about the abandoned mill town on Elk River, and finally, on that summer afternoon, curiousity had prompted me to visit the site. While walking up the valley towards the town I had passed two remote homesteads which showed all the signs of ongoing rural life: cows in the fields, chickens in the yard, a barking dog, neatly stacked rows of firewood, smoke rising from a chimney, and well-tended gardens. The Webbs lived on one of these small farms. Near their homestead was a metal gate which blocked the path to Falk. With Charlie's permission, I later walked that path many times to visit the old townsite, and in 1969, a local historian inspired me to explore the town's past.

Because Falk had flourished and died within the last century, I was able to talk with some of the people who had lived there and experience some of the vitality that had once existed in the town. During these personal visits, people shared some very idealistic memories with me, and it became clear that many of the people I spoke with looked back on their time in Falk as some of their best years. These personal stories, which focus on the lives of the people, convey an emotional element that transcends time. This is the story of Falk's Claim.

Jon Humboldt Gates
July 1983

ORIGINS

Previous page, a redwood forest scene. (COURTESY OF PETER PALMQUIST)

"*This tale begins a long time before man arrived,
and to tell it in its entirety, we must go back to the origins of
Falk's reason for existence . . . the beginnings of
the Great Forests.*"

The Trees
Were Here First

MORE THAN 200 MILLION YEARS AGO, THE FIRST CONE-bearing trees began their evolution during the Mesozoic Era. During this time, the early dinosaurs were just beginning to appear, and it would be 100 million years before the first mammal would roam the earth's surface. Eventually, huge coniferous forests covered the Northern Hemisphere of the planet. These were the Sequoias. It was near the end of the Mesozoic Era, 1.5 million years ago, when the big trees still dominated the land, that a dramatic change took place. The climate began to cool, bringing on the first Ice Age. Vast glaciers formed in the Northern Hemisphere, destroying many of the forests, and forcing the Sequoias to retreat from the advancing glaciers until only three isolated forests remained. In China, a species called the Dawn Redwood escaped the coming of the ice. Across the Pacific Ocean, on the western edge of the North American continent, two other forests survived. One was an inland variety called the Giant Sequoia. These huge trees were located in the southern reaches of the Sierra Nevada Mountain Range.

Farther to the north, stood the last of the three forests, the Coastal Redwood. It was the most impressive of the survivors, covering hundreds of thousands of acres on the continent's central west coast. Some of the redwood trees had life spans of more than 20 centuries, making them one of the oldest life forms on the planet. These ancient trees towered well over three hundred feet in height, and their sprawling bases sometimes measured twenty feet in diameter. For thousands of years the Coastal Redwoods ruled in solitude over fog-shrouded rain forests, their growth recorded only by the trees' inner rings.

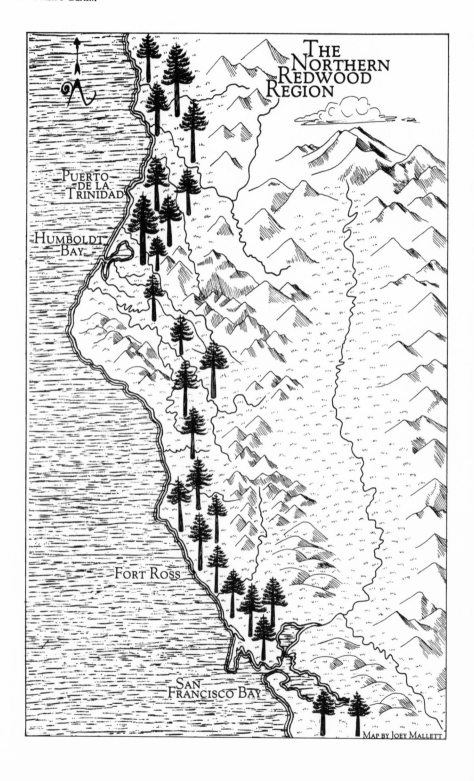

THE
NORTHERN
REDWOOD
REGION

PUERTO
DE LA
TRINIDAD

HUMBOLDT
BAY

FORT ROSS

SAN
FRANCISCO BAY

MAP BY JOEY MALLETT

Geological forces had created a rugged and varied terrain in the redwood region. Along with the dense forests, there were rolling hills, steep mountains, wide river valleys, sandy beaches, rocky sea shores—and in the heart of the region, two large inland bays, located about 250 miles apart. These two inlets were nearly identical in shape, and both were connected to the ocean by narrow channels. The greater of these two bays would later be called San Francisco Bay. It is near the the smaller bay, which lay to the north, that this story takes place.

THE NATIVE PEOPLE

When the Ice Age reached its peak 20,000 years ago, another phenomenon occurred on the continent—the coming of early man. As the glaciers absorbed more and more of the earth's water, the sea levels dropped over 350 feet, exposing land masses that had previously been under water. During this period, Asia was connected to North America at the Bering Straits between Siberia and Alaska. It is believed that early man, in his attempts to survive, crossed this natural land bridge to live and hunt in parts of Alaska which were untouched by the glacial expanse. These were the first human inhabitants of the North American continent.

They remained in Alaska for 8,000 years until an opening in the northern ice corridor allowed them to migrate south in search of new lands to hunt in and dwell upon. As tribes divided and went their separate ways, new languages were formed, and different customs and beliefs evolved. After 100 centuries of migration and development, a mighty Indian culture made up of smaller nations blanketed the continents of the Western Hemisphere.

Linguistic and archeological evidence shows that the ancestral Karoks were the first people to reach the redwood region, settling over 2,000 years ago along the northern boundaries of the great forests. More than 10 centuries would pass before another tribe would enter this remote coastal area. About 900 A.D., a second Indian group, the Wiyot, arrived, followed 200 years later by a third tribe, the Yurok. These two tribes spoke different languages, but were related, and communicated and socialized with each other. The Wiyot had migrated to the coast, far north of San Francisco Bay, and established their settlements around the smaller bay. The Yurok settled a little to the north of the Wiyot, along the ocean's shoreline, and also for several miles inland along the banks of the largest river on the Northcoast.

The other valleys and mountains surrounding the bay were settled around 1300 A.D. by a variety of tribes with a common ancestry. These

A Yurok Indian photographed while spear fishing on the Klamath River in 1893. He was identified as Captain Jack. (COURTESY OF PETER PALMQUIST)

included the Hupa, the Chilula, the Whilkut, the Lassik, the Sinkyone, the Wailaki, the Mattole, and the Tolowa. For hundreds of years these Indians existed in harmony in the redwoods. But in the 16th century—as we record history—major changes were about to take place. Large boats were sighted on the ocean, of a type never before seen by the Indians. These sightings occurred more frequently as time went by but it would be 300 years before the Native American culture would feel the full impact of these "foreign visitors."

DISCOVERIES, CLAIMS AND WAR

The 16th century was a time of European exploration. The explorers were of a different culture than the Native American inhabitants of the redwood coast. They had great ambitions and many clever means for discovery and development which propelled them in their world travels. These first Europeans, in their search for new land and trade routes, seldom came ashore in the treacherous waters along the redwood coast, and pursued virtually no land exploration in the area. Spanish explorers were the first global navigators to chart the unknown sea paths along the West Coast. But there is some evidence that it may have been the English mariner Sir Francis Drake, who in 1579 first sighted the lands where the Wiyot and Yurok lived.

In 1775, two centuries after Drake, two Spaniards provided the first accurate record of Europeans landing on the northern redwood coast. Captains Bruno de Hecta and Juan Bodega anchored their boats in a tiny bay where they were met by canoe loads of Yuroks, who enthusiastically began trading with them. On June 11th, in celebration of the Holy Trinity, the explorers raised a cross on the large headland which formed the bay, naming it Puerto de la Trinidad. Apparently, they never discovered the larger bay just a few miles south.

During the early 19th century, the Russian Empire expanded south from its fur settlements in Alaska. In 1806, an American sea captain working for the Russians discovered the bay the Spaniards had missed earlier. Captain Jonathon Winship named it the Bay of Resanof in honor of Russia's Imperial Chamberlain. Other Russian expeditions followed Winship, attempting to found trading settlements on the redwood coast. In 1808, the Russian sea captain Kuskof anchored in the bay discovered by the Spaniards, but didn't like the location of Puerto de la Trinidad. He noted that the sea otters were depleted and the Yurok village was deserted. Four years later Kuskof found an area better to his liking, and the Russians established the colony of Fort Ross along a rocky section of coastline about 80 miles north of San Francisco Bay. In less than 30 years the fur resources of the area were exhausted and in 1841 the Russians withdrew to their settlements in Alaska.

When the Russians left Fort Ross, rumors began to circulate about a large inland bay north of San Francisco. At this time there were many Americans in California, drawn by the Gold Rush. In 1849, a party of American miners headed by Dr. Josiah Gregg and L.K. Wood, heard of the bay and set out to look for it. They mounted a wintertime expedition to the coast from the headwaters of the Trinity River. On the evening of December 20th, 1849, the expedition re-discovered the elusive harbor, this time naming it Trinity Bay.

But the results of the inland expedition were unknown in San Francisco, and ships continued to sail north in hopes of locating the inlet. Finally, on April 20, 1850, the sailing ship *Laura Virginia* entered the bay, guided by her Second Officer, H. H. Buhne. Someone in this party had a great admiration for Alexander von Humboldt, a German scientist and geographer whose studies and explorations had made him world famous in his own time. The harbor was finally christened with its present name, Humboldt Bay.

After this discovery several more ships arrived, and through federal pre-emption laws, land companies founded four towns around Humboldt Bay. David Buck, a returned member of the Gregg Expedition, started the first of these settlements when he built a single log house and

EUREKA, H.B. 1854.

This early Humboldt Bay scene was painted by a soldier stationed at Fort Humboldt. The Ryan and and McDuff Mill is at the far right. (COURTESY OF CLARKE MUSEUM; PHOTO BY TERESA HENDERSON)

established the town of Bucksport. Following this, Eureka was founded, and also Union Town, which would later be re-named Arcata. The most ambitious city planner at this stage was the Laura Virginia Company, which envisioned a fourth town on the bay, called Humboldt City. The proposed city was laid out and plans were made to build it near the entrance to the bay. A river ran through the center of the town. It had been named the Elk River by members of the Gregg Expedition, who had admired the beauty of this valley where herds of wild elk roamed. But this settlement on Elk River was soon overshadowed by other towns around Humboldt Bay. It would be 30 years before another town would be founded in the valley. And in the intervening period a war with the Indians would be fought for possession of the region's land.

During the unfolding of these industrious years, settlers spread throughout the area, and on May 12, 1853, Humboldt County was formed by an act of the California Legislature. This legislation failed to take into consideration Indian rights to the land. A conflict of cultures was imminent. To head off this possibility, in August of 1851 Colonel Redick McKee, a federal Indian agent, gathered a large herd of cattle and with some 70 army troopers, went north from Sonoma to negotiate treaties with the Indians of the Humboldt Bay region. Colonel McKee offered treaties and land rights to counter native fears over the growing presence of the white men. Nearly all of the Indian groups agreed to the treaties McKee proposed. But when the Colonel submitted his treaties to the federal government, the U. S. Senate refused to ratify them.

What began as two different cultures meeting with mutual curiosity and the desire to trade, grew into open mistrust and violence. In 1852, McKee wrote to the governor complaining about the murder, by white settlers, of 15-20 Indians in the Elk River Valley. The following year, in reaction to increased violence throughout the county, the United States established Fort Humboldt, a military garrison which was located near Bucksport. Local militias were also organized to protect the outlying areas. Despite this, tensions continued to increase, until a dark night in February of 1860, when scores of coastal Indians were massacred by a small, well-organized group of settlers. What followed was five years of open warfare, and the loss of many lives on both sides. In the end, the long established Indian people were forced to make way for the more numerous and aggressive civilization that now claimed the lands. The Native Americans were confined on reservations. This enforced peace brought a new wave of settlers, free to exploit the great wealth of the county—it's magnificent redwood forests.

"THERE'S GOLD IN THEM TREES"

The first settlers who arrived in Humboldt Bay and Trinidad found incredible stands of redwood trees that reached practically to the water's edge. The pioneers who carved their existence from these big trees realized that redwood timber had an enormous potential value. But the sheer size of the redwoods was more than a match for their primitive logging techniques. What was needed to harvest the wealth of the forest was machinery big enough to process that timber.

On February 24, 1852, pioneer lumbermen Ryan and McDuff arrived in Eureka aboard a sidewheel steamer, the *Santa Clara*. Their intention was to build a sawmill on the bay. The two San Francisco men had a channel dug to the shore and ran the steamer aground at what is today 1st and D Streets in Eureka. They constructed a small mill in a building next to the ship, then removed one sidewheel and converted it to a drive shaft to power their sawmill. The ship's quarters were then used as a bunk house by the 35 to 40 mill workers. This Ryan and McDuff operation was the first successful mill in Humboldt County. It ran for ten years until it was destroyed by fire.

Another early pioneer in the timber industry was D. R. Jones. In 1860, he brought the first double circular saw to Humboldt County. Five

Pioneer logging on the Elk River in the late 1870s. During the dry season these huge timbers were cut and piled in the river bed behind a temporary dam. When the rains came, the dam was destroyed, sending an awesome wave of water and logs down the canyon. This was called "flash flood delivery." Note the two men in the picture, dwarfed by the size of the logs.
(COURTESY OF PETER PALMQUIST)

years later, Jones set up a partnership with Captain Buhne, who was then a prominent harbor pilot, and the two men formed the D. R. Jones Company. In 1875, the D. R. Jones Company began to log the heavily-timbered Elk River Valley.

Logging practices in the early days on Elk River left much to be desired, especially if one lived downstream from a logging site. At that time the only way to move logs was by oxen and mule teams, so the loggers felled only the trees which were nearest the river, then cut them into shorter sections with hand saws which measured from 6 to 24 feet in length. The woodsmen usually left behind the lower 20 feet of the tree because these logs were too big to handle. All the work was done in the summer months, so that by fall the river bed was loaded with the sectioned trees. A dam was then constructed downriver of the waiting logs, and as the autumn rains descended, the water level rose until these logs floated freely. The next phase of the operation (and the one that made living downstream somewhat troublesome) was to blow the dam up with high explosives. This sent a flash flood of water and huge timbers cascading down the river. Many of the logs made it all the way down the valley and into the bay, where they were lashed together and towed to the D. R. Jones mill. Quite a few logs, however, ran aground or became tangled in snarls of debris. Jones then sent crews back up river to free the ones that were easily accessible. Those that were too deeply imbedded were not salvaged. As the rains continued to pour throughout the winter, more debris floated downstream and formed log jams around these imbedded snags, which eventually blocked the river and sent it over the banks into the farmlands of the lower valley. This went on for several years before complaints from the farmers forced an end to the flash-flood method of log delivery.

The success of the first lumber operations drew others into the business, and attracted settlers who had been lured to the west by the Gold Rush but had failed to strike it rich at mining. The call from the Northern California coastal forests was "red gold," and even though many claims had already been staked, throngs of people arrived every day, some looking for adventure, some seeking jobs, others hoping to build a future for themselves and their families in the redwood lands. One of those men would soon arrive at Humboldt Bay, to found a logging town in the Elk River Valley that would one day bear his name.

The
Falk Family

THE DISCOVERY OF GOLD IN CALIFORNIA IN 1848
spurred a great westward migration. Just three years prior to the discovery of gold, fewer than 700 Americans had made it to California. But by 1849, California's population had soared to 100,000 people. Most of these "49ers" were Americans, but the cry of "Gold in California" had been heard around the world. Along with the droves of Americans, people arrived from as far as Asia, Australia, South America and Europe. The Americans took one of three routes from the East to California. They could travel overland, which took three months, or they could sail around Cape Horn, which took two months. The fastest—and most expensive way—was via Panama, which took just one month.

During this period of westward migration, Noah Falk, a young Ohioan, heard of the gold in California, but it was yellow gold that had caught his attention, not the "red gold" he was destined to discover. The adventuresome Noah became eager to see what opportunities lay across the continent. Noah's great grandfather, Jacob, had come to Pennsylvania from Germany in 1770, joined the State Militia, and helped to free America in her struggle for independence. Pennsylvania, with its rich, rolling farmlands and forests, was one of the first areas that Europeans settled in this country. It was here that four generations of Falks established their roots. Noah's parents, David and Mary Falk, had nine children during their marriage. Only five of them survived childhood diseases: Noah, Sylvanus, Elijah, Jonas, and their sister, Elizabeth. Noah was the oldest of the children, born on June 11, 1836, in Greenville, Pennsylvania. The family remained at their Greenville home for the next

Noah Falk as a young man, soon after he arrived in California. He, like many others, came west during the Gold Rush, but instead found his fortune in the redwoods. (COURTESY OF HARRY FALK III)

decade before moving to Hancock County, Ohio, where David Falk farmed a government land claim and pursued his trade as a carpenter.

When Noah was eighteen years old, he bid farewell to his family and headed west with "gold fever." It would be more than a decade before the train rails would connect to the West Coast and rather than crossing the wild western part of the country, young Noah traveled south and gained passage on a ship bound for Panama. He traversed the narrow isthmus, boarded a northbound schooner, and on August 19, 1854, sailed through the Golden Gate of California to seek his fortune. After arriving in San Francisco, Noah began to hear of the recent developments in redwood logging that were taking place along the North Coast. These stories led him up to Mendocino County, where he secured a job as a carpenter and helped construct the Albion Mill and Lumber Company. This was Falk's first experience with a sawmill.

Noah later worked his way north to Port Madison, in the Puget Sound area of Washington Territory, where he became a sawyer for the G. A. Meggs Sawmill. Here he acquired a practical knowledge of sawing techniques and general plant procedures, which he later applied to his own mill productions. While living in Port Madison, Noah met a young lady, sixteen-year-old Nancy Brown, the daughter of a fellow worker at the Meggs operation. Originally from Cutter, Maine, Nancy had come to Puget Sound with her family when the mill that employed her father moved from Maine to the Pacific Northwest. A romance blossomed between them, and on August 10, 1862, Nancy and Noah were married.

During the next five years, the couple moved south. Noah ran a 20,000-board-feet-per-day water-operated sawmill near Santa Cruz, California. It was during this time that he began to realize he would like to own his own business. Noah became associated with John Green, who was a baker, and the two men decided to form a partnership. They came to Eureka in April of 1867 with the intention of opening a bakery business. But when Noah arrived in Eureka, lumber baron William Carson found out about his sawmill experience and offered him a job. That offer cut short Noah's career as a baker. Falk worked for the Carson Lumber Company for two years, until he saved enough money to start his own sawmill. In 1869, Noah formed a company called Falk, Chandler and Company and built the Janes Creek Mill, located one mile from Arcata. Powered by a tiny 10-inch by 20-inch engine, this mill became quite successful despite its small size.

The year 1869 was a good year for both Noah and Nancy Falk. Along with their success at the mill, Nancy had a son, Charles. Their first two children had died very young, so the birth of Charles came as a

Saw filers sharpening the double-circular saw blades used in the early redwood mills. The great size of the redwood logs required mills to use two blades, one above the other, to cut through the huge redwoods. Photo is of Scotia Mill in 1889. (COURTESY OF STAN PARKER, PACIFIC LUMBER COMPANY)

blessing. They looked forward to even better fortune in the years ahead. During the next decade, Noah Falk expanded the timber industry in the Arcata area by forming a partnership with Isaac Minor and constructing the Dolly Varden and the Jolly Giant mills. Together, these two mills employed 50 men and produced 45,000 board feet of lumber daily, making them the best mills for their size in the county.

The method of milling used in these operations was a double circular saw, which was developed in the west to deal with the giant trees that were coming into the mills. In the early days of logging, all of the machinery was being built in the east, and the equipment was designed to cut the smaller trees found in that area. The double-circular saw was comprised simply of two singular sawblades, one above the other, enabling the sawyer to cut twice the thickness of a single blade. This saw worked well but still was not able to cut some of the larger redwood logs.

Although Noah had received little formal education, he was a shrewd dealer, and quick to recognize innovation in his field. He was one of the first lumbermen in the area to realize the potential of the band saw, and ventured to put it to use in 1875 when he installed an early prototype in his Dolly Varden mill. This attempt ended in failure, but Falk continued to believe that the band saw would dominate the future in lumber mill operations.

Most of Noah's time was absorbed by his lumber interests, but his inquisitive nature occasionally led him into new areas. During the 1870s, a substantial wheat crop was grown in the county, but few commercial flour mills existed to process the grain. Near their Dolly Varden operation, Falk and his partner Isaac Minor built a small flour mill to grind this wheat. But the project had to be abandoned when housewives complained that the flour from the local wheat strain was too sticky. Noah also made an attempt at gold mining. He and a friend, Henry Jackson, had set up a claim on the Klamath River, but heavy rains swelled the river over its banks during that first winter, and all of their equipment was swept downstream.

Noah turned his attentions back to the logging business which he knew best. In 1875, Minor sold his interests in the two lumber mills, and Noah organized a new company of partners to run the operations. Meanwhile, events back east were soon to give him a partner in his work.

Noah's younger brother, Elijah, had remained in Ohio and, ironically, had acquired part ownership in a small sawmill. Unlike Noah, who had gone west as a single man, Elijah was married. He and his wife, the former Amelia Deabler, had three young sons, William, Charles and Curtis. The ties that bound him to Ohio were strong. But in 1874 their father, David Falk, died at the age of 68 and the ties were loosened a bit. For four years after his father's death Elijah continued working his small mill but began to think of life out in California where his brother Noah lived. Noah had written about the fortunes to be made in the virgin redwood forests and in 1878, Elijah sold his sawmill interests and set out with his family to meet his brother in Northern California. During the next few years, Noah managed his interests in the Janes Creek, Dolly Varden, and Jolly Giant mills, while Elijah learned the skills of a millwright. This was to be a period of great expansion in the timber industry, and many fortunes were made. With the first successes of lumbering in the county, the number of mills and timber interests grew. In 1881, it was estimated that more than 16 billion board feet of harvestable redwood existed in the Humboldt area. The race had begun to see who was going to control it.

LAND GRAB

During the early days of land claims, a settler could purchase 160 acres of virgin redwood timberland for $1.25 an acre, under the condition that he or she was an American citizen and 21 years of age. In the days before immigration laws were passed, one simply went to the courthouse and filed for citizenship, which was granted on the spot. In the 1850s, it was impossible to gain title to most of the timberland in Humboldt County because the U. S. Government had not surveyed the remote forests. One could go into the forest, log the timber and realize the profits, but could not obtain title to the land. This mismanagement was straightened out several years later. In 1862, the Homestead Act was passed, which allowed persons to claim 160-acre land parcels after living on them for six months. This was a tremendous opportunity for a person who had high hopes but little money, as many did in those days. But settlers were not the only ones who desired lands.

News of the vast unclaimed acreage traveled across the country, and big money interests were stirred. The Homestead Act proved ineffective in controlling the redwood lands. Many false claims were made by people who had no intention of settling on the parcels. The government finally moved in and surveyed the forest areas, and in 1878, under the Timber and Stone Act, the first redwood timber lands were made available for sale. The government sold land to any person or association at the rate of $2.50 per acre, in 160-acre lots. This proved to be an excellent deal for the new lumber companies. One acre of virgin redwood averaged 90,000 board feet of lumber, with a going rate of $15 per thousand board feet. A mill could therefore gross $1,350 for every $2.50 invested!

To get land at $2.50 per acre, buyers had to swear, in the presence of two witnesses, that they were not speculating, or planning to sell the deed to someone else. These new guidelines did little to control the land grab that ensued. In some cases, as soon as government land surveyors finished their surveys, insiders arranged for acquaintances to file land claims in the most-desired timber stands. Most of these claimants never saw their land, since it was turned over to a central company. Syndicates and wealthy lumber barons began dealing in timber lands. Some dealt through legal channels, but some did not.

The most notorious case of timberland fraud in Humboldt County, and possibly in the country, began in 1882 and involved the California Redwood Company, which had extensive holdings in the Elk River area. Organized in San Francisco, the California Redwood Company was a consolidation of two small California companies, financed and controlled by a wealthy syndicate from Edinburgh, Scotland. The scheme was to secure over 60,000 acres of prime redwood timberland, for which the Scottish syndicate agreed to pay seven dollars per acre as the land deeds were secured. In order to achieve this goal, the speculators hired surveyors and "timber cruisers" to locate the prime lands of the county, then sent agents to round up 400 persons willing to file at the Federal Land Office for rights to these lands. The company openly advertised for people, going so far as to stop them on the street and solicit sailors from visiting ships to file the false claims. If the sailor was not a citizen, then his citizenship papers were filed for him, even if he was not staying in the country. These false claimants were paid from $5 to $200, depending on the situation.

In this manner, the California Redwood Company managed to secure 64,000 acres of prime redwood timberland. Most of this timber was located between the Del Norte County line and the Mad River. The company also bought other lands outright, until it controlled virtually the whole Freshwater Valley, together with large sections of the Elk River Valley, where it had bought the D. R. Jones Company's holdings. During their peak period of investment in Humboldt County, the California Redwood Company held over 100,000 acres of virgin timberland, two large mills on the bay, another mill in Trinidad, the Trinidad Railway, the Humboldt Logging Railway in Freshwater, and various other holdings. This land grab, however, was too large to keep quiet, and soon Federal land investigators came to Humboldt County following stories of foreign capital fraudulently controlling the richest timberland in the country. Quite a scandal erupted, involving payoffs, Federal land agents, court indictments, the attempted poisoning of a Land Commission agent, and intimidations, both violent and political.

The California Redwood Company survived two years of investigations, during which time the *New York Times* covered the entire story. Finally, in 1897, a judge ruled that the company's actions were fraudulent, and the California Redwood Company was forced to dissolve. Implicated in the fraud were people from the local level up to the brother of the Secretary of the Interior in Washington, D. C.

In the Elk River Valley, the California Redwood Company's land and railroad holdings were legitimate, but after the court's decision, these properties were transferred to the Central Trust Company of New York.

Following the passage of the Timber and Stone Act, all of Humboldt County's redwood lands had been surveyed and drawn into deeds. The stage was set for the orderly development of the timber industry. Once the great redwoods had come nearly to the edge of the bay. But the easy trees had already been felled, and now men turned their gaze inland, where the sounds of the axe and saw had not yet been heard.

Planting
The Seeds

THE ELK RIVER VALLEY, WITH ITS LOW FERTILE LANDS, became one of the first areas in Humboldt County to be settled by ranchers and farmers. The D. R. Jones Company had done some early logging along the river, but thousands of acres of virgin timber still stood in the more remote areas of the valley. These lands, which had been purchased by different lumber companies and partnerships, remained isolated. It was ten miles to the nearest mill, and there existed no means of transport for the logs.

On the South Fork of the Elk River, Judge Cyrus G. Stafford and his partner, Frank Taylor, had invested heavily in several thousand acres of prime timberland. Stafford's original idea had been to sell the trees and hire someone else to log the area for him. But when he surveyed the timber footage, he began to consider the possibility of cutting his own trees, and building a mill on his land to process them. In addition to this, he would also have to deliver the lumber down the valley for shipping. After weighing these possibilities for several months, the judge decided it could be done if the right men were to attempt it. Since Stafford had little knowledge of lumber operations, he contacted an acquaintance of his who was familiar with the lumber industry, Noah Falk.

Noah was living in Arcata at the time and operating his three sawmills. These mills had only a few years of timber resources left, and he was considering what his next move would be when the judge approached him with his plan. Noah was very receptive to the idea, and immediately contacted his brother Elijah to assist in the planned development at Elk River.

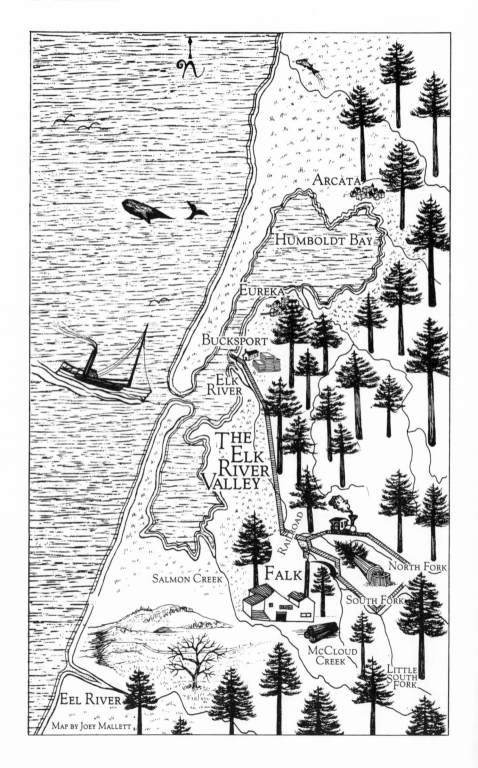

MAP BY JOEY MALLETT

A company needed to be formed to generate capital for the venture. To meet this need the Elk River Lumber Company was incorporated in November 1882, the principal shareholders being Noah Falk, C. G. Stafford, and J. C. Hawley of San Francisco. Because of its isolated location this project involved much more than simply building a mill. The developers had to build a town, complete with homes, bunkhouses, a general store, a cookhouse, a blacksmith shop, a community hall, the mill itself, and a railroad to connect it to Humboldt Bay where a wharf and lumber yard were also to be constructed.

By this time, Elijah had designed and built several mills in the redwood region, and was seasoned for the task of laying out and constructing the Elk River Sawmill. Noah lent his time to organizing a railroad company and securing land to lay the track. Knowing that the mill's success hinged on having a railroad, the principal stockholders had first incorporated the Elk River Railroad Company in October of 1882. The capital stock was valued at $60,000 and 600 shares were issued. Noah Falk retained 300 shares, Judge Stafford 267 shares and J. C. Hawley 30 shares. Two minor shareholders, I. L. Harpster and Benjamin Pendleton held 2 and 1 shares, respectively.

The initial step toward building the railroad was to gain the necessary rights-of-ways from valley residents. During this process, the company was forced to file suit against several landowners who, in opposition to the development, refused to grant the railroad permission to pass. The company eventually gained their rights-of-ways, but lost a considerable amount of time and money in doing so. This was to be the least of their problems. One year later, the company realized it did not have enough money to complete both the railroad and the mill. Even worse, the 25-pound T-iron used on the narrow-gauge rail track was insufficient to withstand the projected loads that were to travel on it. The railroad came to a halt.

Looking for assistance with their project, Falk and Stafford approached the California Redwood Company and the Carson Lumber Company, both of which owned vast amounts of timber in the Elk River Valley. Because it would increase the value of their land holdings, the two companies agreed to take over the railroad construction and ownership, freeing Falk and Company to concentrate on the mill operation and the town that they were building. A new company was formed in July 1884, capitalized at $500,000, with a total of 5,000 shares issued. It was called the Bucksport & Elk River Railroad Company. William Carson held the largest block of shares, with 2,499, while the California Redwood Company held 2,497. The four remaining shares were distributed to A. H. Connick, Joseph Russ, David Evans and Robert Smith. In

The Elk River Mill under construction in 1883. All of the timber used in the mill's construction had to be cut and shaped by hand because there was no mill nearby. Note the dam on the Elk River at the far right of the photograph. (COURTESY OF JUDGE HARRY FALK)

October of that year, the railroad's new owners entered into a contract with Falk's company to complete the railroad with a standard-gauge track, following the line originally planned, and guaranteeing to haul all of the lumber produced by the Elk River Mill.

But the new owners also ran out of funds before the line was completed. So in July 1884 they reincorporated. The California Redwood Company retained all its shares, but Carson's holding was cut in half, to 1,247 shares, while John Dolbeer purchased 1,250 shares. The six outstanding shares were granted to men who held them in the interests of one or the other major shareholders. Essentially, this new company merely represented a reshuffling of interests. Construction resumed and crews continued working up the valley toward the Elk River Lumber Company, eight miles from Bucksport.

Meanwhile, Elijah had completed construction of the mill on schedule, and the plant was operating by 1884. It was an impressive structure. The giant timbers that supported the building were 24 inches square, and some 60 to 80 feet in length. These beams had all been hand hewn with wide-headed axes, and were perfectly straight. A dam had been constructed at a nearby point on the Elk River, incorporating a spillway and a fish ladder for spawning salmon. The first lumber produced by the mill was used to finish these projects and the surrounding buildings. The mill was ready for production, and by June of 1886, the first loads of lumber had left for the wharf in Bucksport. The Elk River Mill was ready for business.

Meanwhile, a third Falk brother had decided to migrate to Elk River to work at the mill. Jonas Falk and his wife, Louise, had apparently remained on the Ohio farm to care for their mother, Mary Falk, in the last years of her life. When she died in 1885, Jonas saw a chance to give his family a fresh start. In 1886, he and Louise came to Elk River, where he took a job as mill foreman, and built a simple two-story house on a knoll overlooking the plant. What Noah had organized, and Elijah had built, Jonas was destined to operate.

Men and Machines

Many of the single men who worked in the mill and woods were housed in bunkhouses and small cabins. Separate bunkhouses were made for the mill and logging crews since the loggers rose earlier to allow time for breakfast and to make their way up into the woods. The mill crews, generally older men, were housed near the mill. Working conditions in the mill were tough by today's standards. The men worked 11½ hours during a normal schedule, but when the mill had a lot of orders, a night shift was also necessary. Since no electricity was available, the night crew worked by the dim illumination of kerosene lamps which were placed throughout the mill. These early mill workers earned wages of $1.50 to $2.00 a day, including board and room. When a man was hired to work in the mill, he usually stayed with his job for a few years at least, so there wasn't much shifting around of employees.

There were some advocates for labor organization around the county who were concerned with working conditions and wages, but the issue lacked focus in the early years of the industry. Phil Flansburg, the first logging superintendent in Falk's woods, was known for his outspoken views in support of organized labor. Flansburg, however, was also a strong supporter of the existing governmental guidelines regarding workers' rights, and opposed any illegal strikes. The work was hard, and the hours long, but the loggers who worked under Flansburg did not entertain the idea of striking because, county-wide, Falk's mill was known to be one of the best camps in which a man could work.

Early log moving in Falk's woods was done entirely by oxen teams. Yoked and harnessed, the teams moved their loads down skidroads towards the millpond. The skidroads were made of small timbers laid across the trails, tended by "skidgreasers" who greased the track, enabling the logs to slide freely. Occasionally the unpredictable oxen became stubborn and refused to move. Irritated "bullwhackers," the men who led the oxen teams, had to encourage them by cursing, prodding, or kicking the poor beasts. These plodding creatures were useful in their time, but mechanical technology began to make the oxen teams a relic of the past.

The 1880s was the decade in which machines entered the woods.

Jonas Falk measuring a fallen giant, perhaps to prove a point to some doubtful relation back east. Jonas managed the mill which his brothers had constructed. (COURTESY OF HARRY FALK III)

Inventions were made that soon affected the entire logging process. The first piece of machinery in Falk's woods was a little locomotive, the Gypsy, brought by Noah from his Dolly Varden Mill, which he had closed in 1885. The engine was used on the logging railroad that ran above the mill. Oxen brought the logs within reach of the Gypsy, which had wire spools mounted on the side. Then the engine, with its short wheel base, would jump up and down as it reeled in its load. The logs were then loaded onto three small rail cars which the Gypsy pulled to the millpond. This engine played a crucial role in the early logging days at Elk River. Years later, Noah reportedly stated that much of his success in the lumber business was owed to the work performed by the little Gypsy.

In 1886 an important logging development, the "steam donkey," made its way into Falk's woods. Invented a few years earlier by pioneer lumberman John Dolbeer, this machine was basically a steam-powered winch on skids, designed to move logs while also possessing the ability to move itself through brush and up steep hillsides. The word passed around camp that one of these machines had been purchased and the oxen would no longer be used. Hiram Frost, whose job it was to clear the brush and build the skidroads on which the logs were moved, did not see how a machine of that size was going to be practical. He and his crew, known as "swampers," talked for days about the "crazy contraption" that was coming to cause havoc in the woods. Finally, the steam donkey was delivered by train. Seeing was believing. The mechanical steam monster winched its way into the back woods with a cloud of smoke and a chugging sound, and once anchored down, it began to reel in the big logs. This introduced a new era in logging Falk's redwoods.

Two years after this first steam donkey arrived, Noah made another major advancement when he installed a Stearns bandsaw in the mill—the first successful installation of its kind in the county. The mill had originally been set up with a double circular saw, but Noah still believed in the bandsaw's superiority. The bandsaw had been designed to ride on wooden wheels but after his unsuccessful experience with the first bandsaw at his Dolly Varden Mill, Noah insisted that the machine have steel wheels. The Stearns Company strongly objected, but Noah would not budge from his demand. The saw was installed as he ordered it, with a 58-foot continuous-loop blade which was 8 inches wide. Other mill owners viewed Noah's idea with skepticism, but when the machine began production, it was a huge success. Practically every major mill in the area followed Noah's lead and changed over to the bandsaw.

Around this time, Noah began shutting down his three Arcata mills. Their timber resources had been exhausted. Noah mothballed their equipment or moved it to other mills where it could be put to better use. He was then free to spend more time at the Elk River plant. Since coming to San Francisco in 1854, Noah had acquired thirty years of experience in the sawmill business and he had many keen insights in avoiding the daily problems that crews encountered. He could be found everywhere—in the woods, in the mill, in the cookhouse, in the offices, and even at the burnpile—always scrutinizing the operations. Many people remember Noah standing out by the mill's burnpile watching the fire. Sometimes at night one could see a silhouette of Noah, leaning on his walking stick, hat pulled low, staring into the coals. No one realy knew what he thought about during those hours, whether he was making sure no wood was wasted, or just letting the flames take him away from his concerns.

One afternoon when the mill crew was getting off work, a newly hired man saw Noah down at the burnpile poking around. Curiosity led the new man to inquire what "that guy" was doing around the burnpile every day? A joking crew member replied that he'd been finding some gold down there. That night a few of the men gathered some brass chips from the machine shop and sprinkled them near the fire. The following evening after work, the new man was out at the burn pile with a kerosene lantern, looking for "gold." He found a couple of the chips and began yelling and running around, but when he showed them to his friends and

A handful of Falk's earliest residents pose for a picture atop a stack of lumber produced by the mill. The white, two-story house in the upper right was the home of Jonas and Louise Falk. The single-story white house below it belonged to Noah and Nancy Falk. Noah later added a second story to the house by jacking the first floor 15 feet off the ground and building a new first floor underneath it! (COURTESY OF JUDGE HARRY FALK)

they started laughing, he knew he had been tricked. He kept a low profile for a few days. Practical jokes were popular around the logging camp.

Noah himself was involved in a somewhat humorous incident, that could have been a lot more serious. It was late one afternoon, during a very busy day, that he was going over figures and costs in the office. He had not had time to sit down for lunch, and was hurrying to complete his work in order to catch the last train from the mill to Bucksport. Noah was so intent in his thoughts that he failed to notice the whistle as the train pulled out. Finishing his work, he rushed to the mill where, to his astonishment, the train was nowhere in sight.

Realizing his predicament, Noah headed for the stables to get a horse and buggy, but along the way he remembered the velocipede in the mill. It was a three-wheeled railroad handcar, in which the rider sat down and pumped a lever back and forth to propel the machine foward. Noah figured the strange machine would get him to town. Once on the tracks, Noah pumped at a good rate of speed, and headed down the valley. When he reached the first trestle, he noticed the bridge siderails were fairly narrow and one of the velocipede's lower tiebars was loose and projecting too far out. While crossing the trestle, Noah reached down to adjust the bar so it would not catch on the bridge but lost his balance. He and the machine toppled over the edge and fell 28 feet into the creek below! Noah was soaking wet and a little shaken, but nonetheless, survived the fall unhurt. Luck had carried him through another incident, and on his coattails rode the future of the company. The Elk River Mill was off to an excellent start. It had direction, capital, labor and resources. All that was needed to ensure success was time.

But on July 21, 1890, the optimism surrounding Falk's mill was suddenly laid to ruin by a single spark. A devastating fire broke out and leveled the mill. When the first outbreak of flames occurred, a worker yelled "Fire!," and the plant was evacuated. Apparently the fire started under the filing room and happened so fast that it was almost an explosion. The entire mill area was covered with a very fine sawdust which acted like gunpowder. In ten minutes the mill was engulfed in flames. The fire was beyond control when Falk contacted Eureka for assistance. An engine from the Torrent Engine Company and her fire-fighters were loaded aboard a special train and rushed to the mill.

Ralph Frost, who was nine years old at the time, viewed the whole event from a hillside and later recorded his impressions of that tragedy in a publication of the Humboldt Historical Society:

> *"My memory goes back to the day I heard the mill whistle blowing just before noon, and we could see the black smoke rolling out of the mill. Then, Mr. Noah Falk, the owner, and a large crew of men in a line reaching to the river, handed five gallon coal-oil cans of water from one to the other, but it was hopeless and the mill burned to the ground. Late that afternoon, they brought the old Eureka handpumper fire engine out on the railroad. We watched that fire burn all day and well into the night. The mill whistle, I guess, was tied down, and it blew until the smoke stacks fell in."*

The engine company labored all night long, and managed to save the dam and several homes near the mill. Along with the mill, 100,000 board feet of lumber burned, making the total loss over $25,000. The people were shocked. For several days all work halted as crews cleaned up the wreckage. At this early stage of development, the company was forced to realize that nature wielded a double-bitted axe; its forest provided opportunities for people, but when the elements of nature were at play, those hard earned fortunes could be easily lost.

Noah called his partners together, to decide the fate of the company and the many people who based their lives around it. The mill was not insured, so the loss was staggering. Also, there remained only a ten-year supply of Stafford's timber to be cut. These two facts caused much discussion among the owners. There had been however, tremendous expenses in laying the railroad to Falk and vast tracts of outlying timber could be purchased when their own supply ran out. Consequently, the owners decided to rebuild immediately.

Elijah, who had been millwrighting on his own, returned to help his brother resurrect the plant. They planned to outfit the mill with more up-to-date equipment, but this was to be done slowly, as financing allowed. Noah went to his idle Janes Creek operation and brought all the machinery from that mill to be used at the Elk River Mill, until newer equipment could be purchased. The project was undertaken with enthusiasm, and by the fall of 1890, Falk's mill was again producing lumber. It had only taken three months to rebuild the plant.

The burning of the mill had been a severe setback for Noah, but his personal fortitude was too great to become undone by such an event. He had demonstrated throughout his life that once he had made a decision to do something, little could stop his momentum.

BLOSSOMS

Previous page, a mature Noah Falk. (COURTESY OF HARRY FALK III)

"I remember one winter when Noah made a big lumber deal and decided to share the profits. He gave everyone who worked for him three extra days' pay as a bonus. He was a fair man in his business dealings."

IRVING WRIGLEY

Cornerstone Arrivals

WITH THE ELK RIVER MILL IN FULL OPERATION AGAIN, life in the camp quickly resumed its spirit. Along with rebuilding the mill and its machinery, more homes were built for the families of the men who worked in Falk's woods. About 30 years had passed since the first pioneers had landed in Humboldt Bay, and for the past decade, a second wave of settlers had been arriving in the region. Many of these newcomers were from timber producing areas in the east, like Maine, Nova Scotia, and New Brunswick. By 1870, more than half of the lumber workers in Humboldt County traced their roots to these areas. Many of these second wave settlers wrote to their friends and relatives in the east, urging them to come west to work in the redwoods. This was how Falk was settled. As the early workers brought their families to the Elk River Valley, they formed the foundations of a community, and began turning a camp into a town.

One of the busiest men in the town of Falk was the blacksmith. A good blacksmith had to be an innovator and a high-caliber craftsman since he was required to fabricate a wide variety of tools and accessories: axe heads, chains, rigging, wheelbarrows, stoves, garden tools, horseshoes, buckets—the blacksmith had to be ready for anything. George Wrigley was the town's first blacksmith. Two of George's uncles had ventured into Humboldt County during the 1870s to try redwood logging, but they had not lasted long. On their return home to the Canadian Maritime Province of New Brunswick, the uncles had some funny stories to tell about living in a place where it took two men three days to chop down one tree.

Their stories created so much interest for George that he decided to travel to Humboldt County himself. The farming in New Brunswick involved long hours and little pay and George wanted to live someplace where he could put some cash in his pocket. In 1884, he and his wife,

George Wrigley and his family were among the first settlers around Falk. George is seated in front with his wife Mary. His son, Winfield Wrigley, who managed the company for many years, is standing at the far left. From the left, the other family members are: Ella Bernice, George II, Mary, Henry, Leila, Theodore, Ruth and Irving. (COURTESY OF RUTH BRAGHETTA)

Mary, and their ten-month-old daughter, Bernice, crossed the United States by passenger train and eventually arrived in Eureka aboard a ship. They settled in the Elk River Valley. George's first job was with a log-floating crew, guiding timbers down the Elk River to Humboldt Bay, where they were lashed into log booms. This was a toned-down version of the earlier method of flash-flood delivery.

Word was out in the valley that the Elk River Lumber Company was looking for a blacksmith. George, who by trade had been a wheelwright, went out to talk with Noah and was immediately hired. One of George's first apprentices in the blacksmith shop was 21-year-old Charles Falk, Noah's son. Determined to learn all aspects of his father's business, Charles made his start with the hammer and forge.

Things in Elk River looked good to the Wrigleys, so they wrote to George's brother James, who still lived in New Brunswick. They told him about the life they had found in Humboldt County. One year later, James Wrigley and his wife, Julia, arrived in the valley, greeted by his brother's family, and their new addition, a one-year-old son, Winfield. James found a job as the conductor of the Bucksport and Elk River Railroad, and twice a day he rode the length of the valley with the first trainloads of lumber.

The growing company needed experienced woodsmen in charge of the more specialized jobs. The "fallers," who were responsible for bringing down the big trees in one piece, were crucial to the success of the woods operation. This job demanded a tremendous amount of precision. If a tree was off its mark, it could be shattered into waste. The chief faller in Falk's woods was handpicked by Noah himself. The man he chose, Herbert Christie, had been a faller at the Dolly Varden Mill before coming to Elk River with Noah. Originally from Nova Scotia, Christie had married Clara Scribner, a young woman from Maine. In search of better prospects, the couple had traveled by train across the country with their four-year-old daughter, Alma, and eventually settled in Arcata, where Herbert got a job at the Dolly Varden Mill.

Herbert and Clara sent news back to their families in Maine of the booming opportunities in Humboldt County. This message resulted in a large group of Christies and Scribners getting together for an expedition to the west. The traveling group consisted of Jet and Ulysses Christie; Benton Scribner, his two sisters, Chloe and Louisa, and a cousin, Lloyd Scribner. Their train ride west was a routine journey, except for one oversight.

Coming from a part of the country where there was little money, this group had to account for every penny spent on their trip, so they carefully packed a large box of food to carry with them on the train. On departure day, Jet Christie was in charge of checking the luggage, and had also volunteered to carry the food box. He was the character of the family, and was always telling jokes and humorous stories while having a bottle close at hand. The kids loved him. When Jet had checked the baggage, however, he was a little intoxicated, and failed to notice that he had also checked the food. The group practically starved going across the country. Jet's name was "mud" for the first part of the trip, but he was soon forgiven because people found it hard to resist his jovial nature. When they arrived in Humboldt County, Ulysses Christie and Chloe Scribner were married, and most of the group settled in the Upper Elk River Valley.

Since at this time no roads connected Humboldt County to the rest of the state, most of the settlers who came to work in the redwoods arrived by ship. For the people with the true pioneering spirit—or no money—several crude pack trails existed that enabled them to reach Humboldt County via land routes. One of the groups who used the overland route was an early Falk family headed by Elisha and Elizabeth Barnes. They were from a well-to-do family in the Woodland area, near Sacramento, but they chose the pioneers' life over the comfort of the valley. The name of Barnes went back many generations to a crusty old

Another "cornerstone" of the Upper Valley community was the Barnes family. Standing from the left: Charles "Pete" Barnes, Loleta Barnes, Ruby Barnes Olson, Robert Browning, Elizabeth Barnes, Vernon Olson, Elisha Barnes. Front row from left: Chester Barnes, Marshal Barnes, Jouko Aho. (COURTESY OF VERNON OLSON)

figure named Shadrack Barnes who fought in the Revolutionary War in North Carolina, and beyond that, to the British Isles, where the family had a coat of arms. Elizabeth's maiden name was Browning, and she was a relative of the English poet, Robert Browning. With their history of moving into the frontiers and their love of adventure, it was only natural that Elisha and Elizabeth packed their wagon and attempted to enter the still isolated northcoast area of California. Leaving culture and wealth behind, the couple and their six children struggled over the hills and rocky terrain, and entered Humboldt County through its rugged south-eastern approach. They spent a brief time camped at the headwaters of the Upper Mad River before continuing on to Kneeland Prairie, where they built a log cabin and lived for two years. The final stop for the family who had "roughed it" for over three years and 300 miles was the Upper Elk River Valley, near the town of Falk.

Another overlander, Evan Rushing, started out from Fort Bragg, California, in 1890. He purchased a mule and provisions and rode north to see what he could find in the Humboldt area. His parents were living in Fort Bragg, where they settled after leaving Pinkleyville, Illinois. At the time, Evan was 18 years old and determined to set out on his own adventure. He traveled the Sonoma Trail which led him over the coastal mountains and along the Eel River Valley to the town of Eureka. His first job was in John Vance's woods, where he became familiar with early logging practices. While working for Vance, Evan met Lloyd Scribner and they shared a cabin together for several years, during which time they became close friends. He and Lloyd often played poker, and occasionally they won as much as $200 in gold on a good night. The two friends combined their gambling earnings, and whenever either won money it was deposited into a special drawer in the cabin. If either man needed money for anything, he was able to borrow from the savings as well. Eventually, Evan met Lloyd's cousin, Louisa, and the couple decided to marry. Before the wedding, the two bunkmates agreed to split their bounty that had been accumulating in the drawer during the past years. It did not take long to divide $3 in gold, which gave the men a grand total of $1.50 each. They thought it was hilarious, considering the many hundreds of dollars that had passed through that drawer during their time together. Several years after Evan and Louisa were married, they moved to Falk and raised a family.

The Woods: A Tough Place To Survive

Life in the town and countryside around Falk was peaceful and simple. The automobile was not yet in production; people rode horses or walked, raised gardens, and chatted on their porches in the evenings. At first glance their lives seemed free from concern, but when the 6:00 a.m. whistles blew to signal the beginning of another workday, the men of Falk entered a hazardous working area. A man never knew whether he would return home that night in one piece or not, but he could not worry about it. It was simply part of the job.

Benton Scribner and his wife, Leila, had settled just north of Falk in a small house and, like the other workers, Benton rode the log train up into the woods in the early morning to begin work. Each day was like the next, and there seemed to be no end to the number of trees that Benton's crew encountered. Benton was a "hook tender," the leader of an eight-man crew which operated the steam donkey that was being used to maneuver the logs around. It was the hook tender's responsibility to attach the rigging onto the logs that were to be moved. This was one of the most crucial jobs in the woods. Only experience told a man how many blocks and tackles were needed to budge one of the tree sections, and he had to be sure that they were firmly anchored as tons of pressure would be applied to them.

Falk woods crew pictured in 1907. The dangerous and physical job of logging demanded team-work that built strong friendships. (COURTESY OF CLARKE MEMORIAL MUSEUM)

Early one morning, after the crew had secured their rigging and cables onto a big log, the steam donkey's winch was engaged and the wire rope pulled tight as it took up the tremendous strain. The log slowly moved up the hill. Suddenly there was a loud crack, and the air was filled with the sound of a whipping steel cable. This was not the first time a cable had snapped loose in a woods operation, but when the men looked around to see if anyone was hurt, Benton Scribner was dead.

One of the most traumatic moments for the families of those workmen was seeing the train come down out of the woods pulling a single flat car with an injured man lying on it, and a friend or two by his side. The families would wonder whose father or husband it was, and in a town the size of Falk, he was certain to be at least an acquaintance.

Benton's widow, Leila, was left to raise their two sons and, though friends helped the family as much as possible, it was a tough situation from which to recover. The following year, Leila opened a postal franchise at her home, and established "Scribner" as the first county-recognized community in the Upper Valley. Since no post office existed in Falk, everyone in the Upper Valley received and sent their mail through Leila's office. Several years later she lost her franchise to the mill's company store, and the town of Scribner was replaced by Falk as the official Upper Valley town.

When a woods death occurred in a family the surviving members faced many hardships. Besides having lost a loved one, there was the new and immediate concern of how to survive, especially if the family was large. Hiram Frost had come to California from Maine in 1880, and married Acascia Gordon from Kneeland Prairie. He worked in several different woods operations in the area before coming to Falk, where he began building logging roads. Hiram's brothers, Dean, Elmer and Philon, joined him in Humboldt County three years after he and Acascia had married. Philon was splitting shakes with Hiram in June of 1897 when a crowbar slipped and dealt Hiram a fatal blow to the head. Ralph was the oldest boy in the family, and the responsibilities shifted to him. He quit school at 16 years of age and washed dishes in the logging camp cookhouse. The cookhouse served three meals a day to more than 70 men, and Ralph worked 15 hours each day, six days a week, receiving room, board, and $16 a month. He gave the earnings to his mother, Acascia, to help look after his six younger brothers and sisters, and with a little help from their three uncles, the Frosts managed to get by.

Aside from the dangers woodsmen encountered in their work, their day-to-day jobs were extremely physical and only the most able-bodied men could tolerate the stress. Thrashing through the brush, dragging cables, rigging, tree jacks, axes and saws, the men climbed up and down the steep overgrown gullies with their loads. There were no coffee breaks. They worked 72-hour weeks and received approximately 14 to 15 cents an hour for their labor, plus meals and a place to stay. The workers' bunk houses were fairly primitive and when a new man hired on he made a mattress of straw, and usually slept under a single blanket. The cabins all had wood heat, so staying warm in the evenings was never a problem. Bathing was somewhat daring, especially on the cold mornings, because the river was the only place to wash. Some of the families had bathtubs in their homes that they filled with heated water from coils in the cookstove. This was considered "high" living.

The men who worked in the woods around the county knew what to expect in these jobs, and conditions were similar all over. Most accepted the hard-working, spartan life of the woodsman. It was a very hearty breed of people who inhabited the town of Falk.

The Trappings Of Civilization

THE RAILROAD LINES WERE THE LIFEBLOOD OF FALK. The little Gypsy locomotive labored above the mill and brought in the log cars from the cutting sites in the woods. After the logs were milled, the Bucksport and Elk River engines delivered the lumber products to the wharf on Humboldt Bay. These two railroads formed the vital links between resource, factory, and market.

After surviving the early frustrations of development, the Bucksport and Elk River Railroad Company had moved rapidly toward achieving its operational goals. One of the prime movers during this construction period was Bernard McManus, the supervisor of the project. McManus had learned about railroad construction in the 1860s while working on the nation's first trans-continental railroad, linking the Pacific Coast to the Atlantic Coast. He and his wife, Kathleen, had emigrated from Ireland to New York during the Great Potato Famine. When Bernard went west with the railroad crews, Kathleen decided on a different route of travel. She boarded a sailing schooner and voyaged around Cape Horn to San Francisco. The couple then re-united and traveled to Humboldt County where they bought a farm in the Elk River Valley. They were already settled in the valley when Noah arrived with his mill and railroad plans. Kathleen was apparently a very determined woman, and she had carried with her an amazing variety of family heirlooms and hardwood furniture during their migrations. Some of this furniture had granite tops and weighed hundreds of pounds, but she had managed to move it half way around the world to Elk River. When the Elk River Valley's railroad tracks were finished, Bernard retired to work his farm. He was constantly reminded of his life's work, since the trains passed back and forth in front of his home every day.

The Bucksport and Elk River Railroad had two locomotives that worked on the line. The No. 1 engine was called the Bucksport, and the No. 2 was known as the Trinidad. The Bucksport was purchased directly from the Baldwin Locomotive Works in 1884. The railroad company had, in fact, bought one large locomotive "kit" complete with parts, assembly blueprints, and operating instructions, and had it delivered by steamship to the Bucksport wharf. It must have been quite a job to put the 23-ton machine together and fire it up for the first time, but the crew managed, and the Bucksport successfully hauled the first loads of lumber in the valley. Because the company needed a second engine to haul extra loads and serve as a back-up unit for the Bucksport, another Baldwin locomotive, the Trinidad, was purchased. This engine was older and smaller, but still a good machine. It had been used at a mill in Trinidad for seven years. Purchase was one thing, but delivery proved to be another. How did one move a 32,000-pound locomotive 30 miles from Trinidad to Elk River when no railroad track existed between the two points? The first idea was to disassemble the engine and send it by ship from the Trinidad wharf to the Bucksport wharf, and then put it back together again.

William Carson, part owner of the railroad company, decided he could avoid all the water transportation difficulties if he were to move the Trinidad overland. Carson contacted a local drayman, Michael McGaraghan, who thought he could move the engine without too much difficulty. The locomotive was loaded aboard a low, sturdy trailer, secured firmly, and McGaraghan's horse teams began to move the awkward load south. At the time, the road between Trinidad and Arcata was complete except for a three mile stretch at Clam Beach. Usually, most travelers simply waited for low tide and ran their wagons along on the wet, hard-packed sand for the three mile distance. McGaraghan intended to do the same.

But apparently, his weight calculations were off. When the wagon reached the wet sand, it sank to its axles. The tide then swept in and covered the engine. There is no record of the commotion that ensued while the men tried to free the locomotive from its sandy graveyard, but the crew did manage and after a dismantling and cleaning, the No. 2, Trinidad, began making its runs through the valley. When not in use, the two engines spent their idle time being repaired or serviced in a train roundhouse that had been constructed at Jones Prairie.

According to some people, the train never pulled a passenger coach, but at least one man, Ralph Frost, claimed to have ridden in one.

In August of 1892 his family was moving back to Falk after living in Eureka for a short time. The Bucksport and Elk River Railroad Company had the right to run their trains over the rails into the Eureka depot, so his mother and brothers and sisters had planned to catch a passenger car from there to Falk. The family arrived at the depot in the late afternoon with a canary in a bird cage, a pet cat, a six-month-old baby, and all their luggage. Everything seemed in order, except that the children were worried about the cat being loose on the train. Someone at the station produced a small box for the canary and the cat was stuffed into the bird cage. When Ralph, the oldest son, boarded the passenger coach holding the cage, he remembered thinking that his family must have looked a bit odd to the people they met that afternoon. When they reached Falk, with the cat and bird in good health, the travelers were glad to be home again.

Although there was a road and several trails which also gave access to the Upper Valley, it was the success of the railway and the mill that ensured the town's future. But while these two enterprises provided the basis for economic security, it was the people's own initiatives that began to develop a community in the Elk River Valley.

Reading, Writing And Religion

The first common interest among the families of Falk was to raise a schoolhouse for their children. At the request of the townspeople, the county school board formed the Jones Prairie School District, named for D. R. Jones, who had first logged the Elk River Valley. The county also built a one-room schoolhouse in nearby Scribner. The first classes at the Jones Prairie Elementary School began in February 1887.

The school's facilities were fairly primitive in the first months of classes. The furniture and desks arrived some weeks after the students, and the outhouses were not completed until late springtime. Whether these conditions created some hardships is difficult to say, but the first teacher, Miss A. L. Doe, lasted only one month. The early instructors strongly emphasized basic reading, writing, and arithmetic skills, but they also found time to encourage creativity. The school acquired a foot-pump organ to accompany singing lessons, which, next to recess, proved to be one of the children's most enthusiastic gatherings.

When a youngster attended Jones Prairie School, two attributes were required by the instructor: discipline and responsibility. Teachers expected an orderly class, and those who had a problem remembering that would not forget being sent outside to cut their own whipping stick. The youngsters were also responsible for several classroom chores. Cords

Alice Wrigley and her class from the Jones Prairie Elementary School. Alice taught at the school for over 50 years and later retired as principal. In 1910, a second schoolhouse was built next to the original one. The two schools had a combined attendance of nearly 100 children. (COURTESY OF RUTH BRAGHETTA)

of firewood were delivered by flat car from the mill to a landing near the school. After a hired man had split the wood, the children stacked it in a shed next to the schoolhouse. The older boys made kindling and helped to keep the woodstove burning, while the younger children took turns cleaning the open trough that fed spring water into the school's water tank. The kids were used to chores because, most likely, they were required to do them at home.

Occasionally, a class outing was organized for the youngsters. These field trips generated a tremendous amount of excitement. On Columbus Day of 1892, Myron Young, the new schoolteacher, took the 60 students of Jones Prairie School to join the Elk River and Bucksport Schools in a celebration. The children were loaded aboard flat cars, and a railroad locomotive delivered them down the valley to the Elk River School for an afternoon of games, picnicing and photos. The children enjoyed these picnics and outings, even if it was only five or six miles down the road.

Classes at the school went only as far as the eighth grade level, at which point the boys had to make an important decision. Their choice was to attend four years of high school in Eureka and commute over 20

The First Congregational Church after a Sunday service. The Parson's cottage is at the lower left.
(COURTESY OF MR. AND MRS. JOE BARLOW)

miles every day by stagecoach, or go to work in the woods and start earning a paycheck. The 13 and 14-year-olds who decided to go to work wheeled firewood around to the cookhouses and shop stoves, or became a "whistle punk" in the logging operation. That job entailed blowing a whistle to tell the donkey engineer when to start or stop the winches. But no matter what the job was, if a young person quit school, they usually never returned.

The schoolhouse became a central gathering hall for the Upper Valley residents since it was the only large building that wasn't owned by the company where a crowd could comfortably get together for discussions or meetings. It also became election headquarters for local, state, and federal pollings.

Although many of the families that settled the Upper Valley desired to attend Sunday services, no church existed within miles, so the choice was a long buggy ride, or reading the Bible at home. Several people contacted the Congregational Church of Eureka and requested

The Herbert Christie Family was among the prime movers behind creation of Falk's church. Top row from left: Glenna, Alma, Gladys. Bottom row from left: Ruth, Herbert, Clara, Ivan. (COURTESY OF RUTH JOHNSON)

that an evening service be performed. In September of 1894, the Eureka church sent Reverend G. Griffiths to lead the first gathering at the Jones Prairie Schoolhouse. Rev. Griffiths urged the assembly to continue requesting regular services. A few months later, Reverend. G. A. Jaspar came to the Upper Valley and conducted his first service. After several Sunday services, a meeting was called and the Congregational Church system was explained. With a show of hands, the group decided in favor of forming a Congregational Church. The people elected a committee to accompany Reverends Griffiths and Jaspar to a county-wide meeting of church representatives. Finally, their request for a church was granted in March of 1895, and at their first official meeting, a ten-point constitution was written and officers were elected. The First Congregational Church of Elk River was born.

Several townspeople devoted a good deal of energy toward organizing the church in its early days. Herbert Christie gave much of his time, along with George and James Wrigley and Jonas Falk. Mr. C. V. Wharton donated nearly an acre of land for the church to be built on, and through the Elk River Lumber Company, Noah Falk donated all of the materials and lumber to raise the structure. The church purchased books, benches and an organ, and paid J. M. Pluke to paint and paper the building. They probably hired Pluke because when it rained, he let the high school students from Falk stand under his paint store's roof in Eureka while they waited for the stagecoach to take them home at night.

Each Sunday, before church services were held, Herbert Christie directed over 60 children in Sunday School classes. The youngsters arrived in the early morning, and gathered in one corner for their Bible lessons and singing. Herbert gave a lot of himself to community interests, and with what little time he had away from his 72-hour-a-week job, he and Louisa also raised a family of five children.

The town of Falk had achieved a full circle in its development when the church was completed. For the townspeople, who founded their principles around simple teachings, the town of Falk provided a livelihood, a home for their families and an education for their children. Now, they had a place in which to give thanks for it all.

The New Century

THE ADVENT OF THE TWENTIETH CENTURY BROUGHT many new challenges to the Elk River Mill and Lumber Company. Men, machines, and land holdings were all affected during this period of growth. By 1900, the mill had cut all of Stafford's trees in the McCloud Creek Basin and for business to continue the company was forced to expand their timber properties. As a result, in 1901 Noah purchased 3,000 acres of harvestable redwood from the Excelsior Company. This land, the same tract that had provided the incentive to rebuild the mill after the fire of 1890, was located on Elk River's Upper South Fork, several miles above the town. Because it was an isolated area, this land acquisition created new difficulties.

Extending the logging railroad was essential. The line would have to climb more than five miles up a narrow, winding canyon, and nearly a dozen large wooden trestles would have to be built to cross the difficult terrain. It would also be necessary to build a logging camp to house the single men closer to the new location. This camp needed crew cabins, a cookhouse, a filer's shop, and a blacksmith shop. The project took two years to complete. In 1903, the first harvest of logs from the upper canyon were ready to be sent down to the mill.

The Gypsy engine, which had been hauling the company's logs for 17 years, still ran well, but the new railroad was too steep for the little machine and it was retired. The Gypsy had developed quite a personality over the years, and for some people on the Upper Elk River the Gypsy became the town's mascot. Taking it's place was a new 20-ton Baldwin locomotive. This engine was more powerful and could easily deliver a dozen log cars down the canyon in one trip. The Baldwin immediately

Posing aboard the Gypsy engine are (l. to r.) Albert McManus, Ted Wrigley and Elisha Barnes. The Gypsy, known as the No. 1, was the first machine to be introduced into the Elk River Valley's logging industry. (COURTESY OF EVELYN OBERDORF)

became the workhorse of Falk's railroad system. But this wasn't the end of the line for the Gypsy—it now did millpond chores, grocery deliveries, and camp errands. Several years after the Baldwin's arrival, the little Gypsy was involved in a mishap which furthered its reputation as a source of nuts and bolts humor.

The incident occurred near the millpond. Albert McManus was running the Gypsy up the tracks, while at the same time, Jesse Barnes was coming down from the woods with the new Baldwin locomotive. Jesse's engine had the crew boxcar in front of it, so his visibility was limited. When Albert saw the Baldwin coming he quickly slowed down, thinking that Jesse would do the same. But Jesse, who knew the line blindfolded, had his head out the window and was looking at his cows that were grazing in a nearby meadow.

Suddenly, Jesse saw the Gypsy and slammed on the brakes. Albert, foreseeing an inevitable smashup, threw the Gypsy into reverse at full throttle, and leapt from the cab just before impact. The collision was so slight that neither of the engines, nor the boxcar, left the rails. But the Gypsy, in full reverse, took off backwards down the track with Albert chasing it on foot. He ran his fastest sprint but could not catch the runaway machine. The Gypsy charged through the town of Falk and headed for Bucksport!

A startled onlooker burst into the mill office and yelled that the Gypsy had just raced through town with nobody on it. Several people joined the chase after the freewheeling engine. Two miles out of town, the Gypsy was found at an abrupt stop. It was standing on end with its back wheels run up onto a flatcar that had been left on a siding. Had it not been for the siding's open switch, the Gypsy's career most likely would have terminated at the end of the rail line in Humboldt Bay. The engine had sustained minor damages, but it was soon repaired and back in service.

Replacing machinery and resources was a constant concern for the company. But these were only some of the changes that had occurred around the town of Falk. The townspeople changed as well, through retirement, injuries, firings, and other events that opened the doors for new people to enter. During this period of expansion, Charles Falk assumed his first leadership job, taking over as mill foreman from his Uncle Jonas. In the 12 years that Charles had worked for his father's company, he had gained practical experience in virtually every area of operation, from the blacksmith shop, to the woods, and now he directed the mill's operations.

When Noah organized the Elk River Mill, his close friend and longtime business manager, Irving Harpster, had overseen the office details and kept accurate accounts of all the commercial dealings. But the growing business that Falk's mill was experiencing began to flood the office with paperwork, and it became necessary to hire a full-time bookkeeper. There could not have been a more suitable candidate for this job than Winfield Wrigley. His father had been the blacksmith at Falk for 19 years, so young Winfield's roots were in the Upper Valley. Winfield knew most of the people in Falk, had studied business in school, and was familiar with the lumber company's operations. He was quickly chosen for the job and eased into this new position. But several years later, what began as an unbalanced payroll led Winfield to uncover a case of fraud within the company.

After a bit of detective work in the woods operations, he discovered that the woods boss had a dead man on the payroll, and had been collecting double wages for quite some time. When Winfield reported his finding, Noah, who was usually calm and collected, became furious, and immediately summoned the offender. Noah stormed around the office, his voice cracking as he cursed the man, and fired him on the spot. After his tirade had passed, Winfield was promoted to Secretary/Manager, and Noah set out to find a new woods boss and another bookkeeper for the company.

In 1905, two years after Winfield had been hired at the mill office, his father, George, almost died as a result of food poisoning. The incident occurred at his home in Eureka, where the family had lived for the past five years. Though he survived, his health was severely impaired. Before this misfortune, George had demonstrated his vigor by rising before dawn each morning to walk across Eureka and then pedal eight miles out the Elk River Valley on a track bike, which had three flanged wheels and was designed to be pedaled on a railroad track. He followed this warmup with a full day's labor, and returned home each evening in the same manner as he had arrived. George had maintained this routine six days a week, but after 20 years of blacksmithing, the food poisoning incident forced him to give up his work. The Wrigley family moved back to Elk River and settled on their farm near Falk, where George rested and attempted to regain his health.

Noah could waste no time in replacing George, whose departure was sad and sudden. But the job was an integral part of the operation and within a week the bellows were in full flame again when Jack Miller became the new blacksmith. Jack, who had lived in Humboldt County for 12 years, had worked as a smith in several of the area's logging camps. He had first landed in Eureka by ship in 1893, following a three-year journey by horse and mule from his native land of Texas. Jack's first day in Eureka caused him to briefly wonder why he had ever chosen to leave the dry inland prairies of his native state. His ship had weathered a violent winter storm on its passage to Humboldt Bay, and for three days the schooner pitched and struggled its way up the coast. Jack had never been on the ocean before and was miserable with sea sickness. Despite the torrential rains that were falling when his ship entered the bay, he was more than ready to set foot on solid ground again. Groggy-eyed, Jack stepped off a wooden sidewalk in downtown Eureka—and sank to his knees in a mud hole. But he had come too far to be so easily discouraged. Despite his wobbly knees, the rain, and the mud, Jack decided that he had found a home.

During one of Miller's first days on the job at the Elk River Mill and Lumber Company, Noah and Charles stopped by the blacksmith's shop to see how things were going. Jack had been preparing a bearing when they arrived, and as he poured the material into the damp mold, Noah leaned over to inspect the job. "Get away from there!," Jack yelled, "She's going to blow!"

"No way," said Noah, who had barely finished his words when the metallic liquid blew out and covered his hat, jacket and beard in molten metal. Jack rushed to Noah and brushed the hot substance from his

The blacksmith shop at Falk. Standing from left: Evan Rushing, blacksmith Jack Miller and filer Simon Fraser. The blacksmith was one of the most important craftsmen in the logging operation. (COURTESY OF EVADNE BRADLEY)

clothes and beard, scalding his hands in the process. Noah was not hurt, but his jacket was badly stained and his beard was singed. Noah wore that same stained jacket every workday for the next seven years.

CHANGING TIMES

During the early 1900s, industrialized countries everywhere began to experience historic upheavals and transformations. Most of the events that happened in the outside world held little or no concern for the people in Humboldt County. But a growing union movement, supported by the theories of socialism, began to penetrate even into the isolated woods and mills of Humboldt County.

The International Brotherhood of Woodsmen and Sawmill Workers had begun to organize in the county to improve wages and working conditions for its membership. They had watched economic conditions improve for others and seen fortunes made in the lumber business. And they wanted more of the wealth.

The union had set forth a list of demands which included higher wages and the re-establishment of free board. Only four mills in the county granted these conditions without complaint. The Elk River Mill and Lumber Company was one of them. Falk had never changed his free board policy, and needed only to comply with the increased wage demand, which was granted. But many of the larger mills would not bend to the demands, and on May 1, 1907, 2,650 men walked off their jobs. This was out of an estimated work force of 3,000 timber employees. Lumber production in the county was practically halted.

This strike had no effect on the crews at Falk. Noah had always managed his businesses through personal contact with those whom he employed. His employees had no need for a union. It was not uncommon to find Noah at the mill or up in the woods having lunch with the logging crew, and he was easily accessible to anyone who needed a word with him. This direct contact had a positive pay-off, and the Elk River Lumber Company was never disrupted by the labor-management confrontations that crippled others in the industry.

Even though the town of Falk could solve its local concerns, forces were at play in the world which affected the town. In 1914, the countries of Europe went to war and within a year even the Humboldt economy began to suffer. With the demand for lumber products dropping off and a lack of available shipping to move the lumber, many mills closed their operations. The Elk River Sawmill was among them. The people of Falk lived patiently in sight of the shutdown mill in hopes of a revived economy. Nature, which had seen the town carved out of a virgin redwood forest, was now testing the little community's perseverance.

In May of 1915, several months after the mill's closure, Noah made a trip to San Francisco to attend a lumbermen's industrial conference and to visit the Panama-Pacific International Exposition, held to celebrate the opening of the Panama Canal and the establishment of a shorter sea route between the east and west coasts. When he returned home, newsmen met Noah and Nancy as they stepped off the ship. He had very positive news about the recovery of the lumber industry, and told reporters that his mill would reopen within a week. It had been shut down for four months, and the woods operation for eight months.

The depressed economy which caused Noah so much concern had also prompted a conversion in his brother, Elijah. He had followed Republican politics for much of his life, but in 1906, he began to study socialism. When Elijah retired from the lumber industry, he devoted himself to political interests, and in 1915, campaigned for the position of Mayor of Eureka on the Socialist Party platform. He was very popular among working people but the business community and the press were united against him. During the campaign his opponents received extensive coverage in the local papers in support of their viewpoints. Only one article appeared with mention of Elijah, and that was a derogatory editorial concerning the evils of socialism. Nevertheless, on election day Elijah bettered his nearest opponent, L. F. Puter, by the narrow margin of three votes. The opposition demanded a re-count. But the three-vote margin held. On July 6, 1915, Elijah Falk became Eureka's first Socialist Mayor. During the same election, Jonas' son, Harry, was elected Police Judge.

Preceding his political career, Elijah had served the timber industry for over 25 years. His achievements in mill construction had become recognized state-wide, and he had built more mills than any other man in the redwood region. He had also revolutionized the shingle industry when he designed and built the first shingle drying plant at Bucksport. In 1904, author J. M. Guinn wrote in his *History of the State of California"*:

> *"In connection with the Bucksport Shingle Mill Mr. Falk built a drying plant at the mill, which was the only shingle dryer of the kind in the country. All of the shingles were dried before shipping and their weight was thus reduced ⅔, which made a great saving in freight. It took several days to dry the shingles, which were dried at the rate of 120,000 per day. The kiln held 1,428,000 shingles at one time, as 204 cars, each holding 7,000 shingles, were put into the kiln at once."*

Over the years, Noah and Elijah were instrumental in developing the area's resources and economy, and had influenced both the timber industry and the public with their innovative ideas. They had started a railroad, constructed, owned or operated over 13 mills in Humboldt County, been successful in politics, and most importantly, raised a town that bore their family name. These were the beginnings.

Elijah Falk, Eureka's first Socialist Mayor. Elijah designed and constructed eight mills in Humboldt County: the Bucksport Shingle-mill, the C. K. James Sawmill, the Carson Shinglemill, the Elk River Sawmill, the Harpst Shinglemill, the Shipyard Sawmill, the Vance Mill and the Warren Miner Sawmill and Shinglemill. He also built the Hines Sawmill in Santa Cruz County. (Private photo collection)

The Company Matures

By 1912 the Bucksport and Elk River Railroad Company had been in business for 26 years. The line had steadily improved during this time and now employed a number of men from the Upper Valley. William Carson, who owned half the railroad's stock and was the line's principal director, died in 1912, at which time J. Milton Carson inherited his interests and became president of the railroad. The other half of the railroad's ownership had been previously restructured when two San Francisco men, J. R. Hanify and Albert Hooper, bought out the entire stock of the scandal-ridden California Redwood Company. The title transfers were for the most part a paper shuffle, and were little noticed by the railroad crews since their work orders and paychecks continued to arrive regularly. In 1913, however, work on the railroad was interrupted by the death of a long-time employee who was one of the most respected men in the valley.

James Wrigley had started with the Bucksport and Elk River Railroad Company as a conductor and later rose to superintendent of the operation. He had ridden the rails of the valley for 25 years, and was leaving behind his wife, Julia, and seven children. A story has been told about James which sums up this man's gentle disposition and patient outlook.

One Sunday, just after lunch, James and Julia sat in the quiet of their home that stood beside the Elk River. Across the river lived their neighbor, Frank Gossi, who worked in the mill. It was the weekend; no trains were running, no mill whistles were bellowing in the valley, and Frank was eagerly working on his land, clearing rocks and stumps in

Clarence and Nellie Weatherby. Because so many family members worked on the railroad, the Weatherby name and the Bucksport and Elk River Railway were almost synonymous to people in the valley. (COURTESY OF RUSSEL WEATHERBY)

order to have a good crop planting field. One exceptionally large stump was quite a problem to remove. Frustrated, Frank stomped off to his barn and came back with a bundle of dynamite. He packed the charge into the old stump, lit the fuse, and ran for cover. There was a terrific explosion that sent debris and earth flying in every direction. He had used far too much explosive for the job, and literally disintegrated the snag. One of the larger pieces was blown high into the air and across the river. It came down with a crash against the side of the Wrigley house and smashed their rain barrel. Several seconds later the front door flew open, James stalked out, pulled his hat down on his head and crossed the footbridge that led over the river to Frank's place. "Oh, oh," Frank thought. But James walked up to him and said in an even voice, "Frank, for God's sake, will you please be more careful with your powder?" He then turned around and went back to his home and nothing more was said.

The passing of James Wrigley left the Bucksport and Elk River Railroad in need of a new superintendent to manage the line's schedule. Carson and his co-directors chose Clarence Weatherby. Clarence had lived in the valley for 17 years, and had been an engineer with the company for most of that time. Having worked closely over the years with James, he was familiar with the superintendent's duties. The company also hired Jack Weatherby, Clarence's brother, to fill the vacated engineer's job. To further complicate matters, Clarence's son, Russell, became the fireman on his Uncle Jack's locomotive. During the next decade the name Weatherby and the Bucksport and Elk River Railroad became almost synonomous in most people's minds.

The train trip from Falk to Bucksport was a routine run, interrupted only by an occasional mechanical breakdown. But because there were many livestock in the valley, the engineers had to watch carefully for stray cattle on the tracks. One afternoon on the return run, Jack's locomotive had built up a head of steam and the empty train was running light. He had just rounded a tight bend near Engles Corner when a bull walked onto the tracks. There was no time to stop the speeding train, and the large animal was struck full-force by the locomotive. The impact was so great that the engine and five flatcars were derailed from the tracks. Although Jack was unhurt, he was a bit shaken. The accident resulted in only minor damage to the train, but the bull had become a victim of the metal cowcatcher.

Though an incident like the Engles Corner derailment could create some local valley news, it was insignificant compared to the Northwestern Pacific Railroad's announcement in 1914 that the tracks were complete between Eureka and San Francisco. Humboldt County had finally accomplished a commercial, land-based connection to the outside world. While steamers loaded at the Bucksport wharf continued to ship most of the lumber that left Falk, the completed rail connection to the south opened a new dimension to the Elk River Mill's shipping procedures. In addition to loading the daily lumber trains for the Bucksport wharf, mill crews also began loading cars to be shipped to distant points. These cars were delivered down the valley and parked on a siding where the Northwestern Pacific locomotives picked them up for their run south. The company shipped their products via the Canadian Pacific, New York Central, Santa Fe, and Missouri Pacific. Virtually every corner of the country began to receive redwood milled at Falk.

THE COMPANY OFFICE

The mill was now producing 40,000 board feet of lumber daily and the company was swamped with work orders, sales orders, and a bulging payroll of well over 100 employees. Noah and Nancy had moved to a permanent residence in Arcata, but Noah continued as president of the company and still directed its operations. During the past decade he had gained confidence in Winfield Wrigley's abilities as manager, so on his days of absence, Noah left Winfield in charge. The company offices, located in one end of the general store, became the scene of many little dramas, not all of which were related to business. Vernon Olson, who was hired as the company's bookkeeper in 1913, discovered that his duties often involved much more than the simple concerns of balanced ledgers. One of his first predicaments came only weeks after he began working in the office.

It was late afternoon and the lumber train was pulling out of town for its run to Bucksport. Vernon was at his desk, quietly going over the books when a commotion erupted outside the general store. People began to yell for help, calling out that someone had been run over by the train. Vernon sprang from his desk and ran out of the office toward the crowd, and discovered a poor drunk with one of his legs missing. Vernon hurriedly took off his own necktie and applied it as a tourniquet to the injured man's leg, then ran to borrow a friend's Model-T and raced the sobering man to the Union Labor Hospital in Eureka where the injury was treated. "Funny thing about it," Vern said, "the next few times I saw the fella, he kept complaining that his toes itched—on the leg where he didn't have any toes."

Before he came to Falk, Vernon had worked at his father's retail lumber yard in Loleta, and had also attended business school. Noah had heard of him through acquaintances and personally offered the position of company bookkeeper to him. Vernon accepted and moved to the town, taking a small room above the cookhouse. His starting wage was $35 a month, plus board. Little had he realized that his move to Elk River would bring more than a new job and paycheck into his life. During his first year at the mill, Vernon met a young lady who worked in the upper cookhouse, Ruby Barnes. The following year, they decided to get married. Ruby was only 16 years old, and because she and Vern feared her family might disapprove of their marriage, they stole away in the early hours of the morning with the help of her cousin, and boarded the steamer *Corona* which was sailing to San Francisco. The next day they were married. This elopement occurred at a time when the company was very busy and Vernon had been able to take only a four-day leave. Upon their return, the newlyweds moved into a small house they had rented and Vernon went back to his desk to catch up on his records and ledgers.

The busiest time around the office was always the first Saturday of every month—payday. The bookkeeper and his assistant issued the workers their monthly checks. Vernon's assistant was William "Red" Braghetta, who had worked in the general store and come to know many of the townspeople, including Vernon. When the time came to hire an office assistant, "Red" was chosen.

Payday invariably created a sense of anticipation among the work-ers. When the mill whistle signaled the end of another week's labor, the men swarmed to the office for their checks. The mill crew was always paid first, since the woodsmen had to ride the log train down the canyon and arrived in town about half an hour later than everyone else. The workers called out their names while filing past a small office window to receive their pay. "Red" wrote the checks and Vernon recorded and issued them.

Occasionally at night when Vernon (or "Ole," as some of the workers called him) was sitting at his desk, he would hear a tapping at the window, "Ole, Ole, I need ten bucks." Ole would reply, "But you don't have it coming." The voice from the window would respond, "I know, I know, but I need it." Vernon knew most of the men personally and always trusted that they would not cheat him, so he advanced the few dollars needed. In the 19 years that he worked at the mill, the only money he lost on good faith was $20 to a man who skipped town.

During these years, the Elk River Mill and Lumber Company's office also became responsible for an independent operation that worked on their land. Hiram Thrapp produced redwood ties with a small steam-powered mill that sat up on the hill above town. The Elk River Mill bought all the ties he produced, either to use on their own railroad, or for resale to other lines. They also supplied Thrapp with some operating materials, and issued the pay to his hired help. Hiram sent his men's wage cards down to the office, and Vernon took care of the checks.

On one of those routine paydays a huge man that worked for Hiram accused Ole of shorting his pay, even though it matched up with the wage card his boss had sent down. The man was beyond reason and demanded extra money or he was going to come into the office and tear things apart, bookkeeper included. Ole wasn't a big man but he was fiesty and was not going to be intimidated. He grabbed a wire rope paperweight and threatened to defend himself with it. The irate worker's bark proved worse than his bite, and he left the office with no further argument.

Having the independent woods operation on company land caused special complications. The company allowed Thrapp to cut timber, but the only claim he had was to his profits, and not to the land or property. When Noah first started the company in 1882, free lots of land were given to people who planned to work at Falk. They were also given free wood, paint, and wallpaper to construct their homes and maintain them. After several early incidents in which people tried to claim homestead rights to their land parcels, the company changed the policy. One could still buy or sell a house, but they could not own the land it sat on. A token fee of 50 cents a month land rent was charged to those who owned

Enjoying a day at Clam Beach. From left, Norton Steenfott, Ethel Steenfott, Ruby Miller, Charles Steenfott, Jack Miller. Wayne Miller is seated, holding the dog. Bottom, from left, Evadne Miller, Nathalie Steenfott, Nedra Steenfott. (COURTESY OF NORTON STEENFOTT)

homes on mill property. Still, a minor land controversy did arise in the later years with regard to Charles Steenfott's home.

During the summer months flies invaded the homes and swarmed around the ceilings, making an annoying sound. Fly paper was not enough to combat the pests, but there was a technique that was effective. Ethel Steenfott would light a newspaper on fire and singe the flies' wings off, then sweep them up. One day when Ethel was away from home, her daughter decided to help rid the house of flies. She ran around the house burning the fly wings, which was great fun, until she dropped the newspaper on the wood floor. Instead of stamping out the fire, the frightened youngster ran for help. When people arrived the home was engulfed in flames, and burned to the ground. Because Mr. Steenfott was employed as the company millright, he was given free wood to build his family a new home, and within two months a fine house was ready to be occupied.

Later in the year Steenfott was hired at the Mercer-Fraser Company in Eureka, but he and his family continued to live in their new home at Falk. Winfield Wrigley discovered the situation, and since the house was on company property, informed Steenfott that he had to move. Charles said no, that he had built the house and paid land rent, and he wanted to live there. He went into Eureka for legal advice on his rights and that's when matters got even more complicated. His house was actually built on land which belonged to Dolbeer & Carson Lumber Company, and was not part of Falk's land at all. The Dolbeer & Carson Company representative contacted the Elk River Mill and became quite irate over the fact that it had given Steenfott the right to build on their land. It was a three-way squabble, but Charles came out well in the end. He received $1,500 in cash, and the Elk River Lumber Company boxed and shipped all of the family belongings to their new home in Eureka.

The Steenfott case was only one of several instances of controversy which developed over the years between the company and the townspeople. The company office usually ended up settling these disagreements involving their policies, but not everyone enjoyed the outcome, including the company. For the most part, though, the offices were the central point for business, and in a booming mill town like Falk, that business was turning logs into lumber.

Logging Falk's Redwoods

HARVESTING THE REDWOOD TIMBERS WAS A TEAM EF-
fort and each man's job was a vital link to the next man's. Many of these
woodsmen had worked with their partners and crews for years and had
established an undeclared brotherhood between themselves. It was with
the strength of this bond that they entered the woods.

The forests of the Elk River Valley were vast, but each year as more
trees were cut, Falk's logging crews moved further into the back country.
Several small camps were constructed in the more remote areas, but they
were essentially crew quarters, and relied on the main camp for their
services. Most of the men who lived in these upper camps were single; the
men with families lived in town. At five o'clock each morning the
woodsmen from town gathered at the locomotive barn to board the crew
train for their five-mile commute up the canyon. It was a 30-minute ride
in an open boxcar to the main camp where the men were served a huge
breakfast before beginning work.

The woods crew was divided into two groups, headed by tree
fallers, known as "choppers," and machine operators. When the logging
operation entered an uncut forest, the choppers were the first men into
the woods. Their job was to cut the logs which the machine operators
and their crews were to haul out the following year. Working in teams of
two, these men used double-bitted axes and long crosscut saws to bring
down the giants. Since the trunks of these trees were considerably wider
at their base (and the tree's grain structure was harder to cut through
than at higher levels), the fallers worked on scaffolds called springboards,
attached five to ten feet up the tree. The men sometimes spent as much as

Pausing for a moment during the workday is Herbert Christie, left, chief faller in Falk's woods. Woods Boss Jim Copeland is seated at left, with Noah Falk next to him. The man with the axe is unidentified. All four men are actually several feet off the ground as they are posed on "spring boards" which the fallers worked on. (COURTESY OF RUTH JOHNSON)

three days on these springboards, laboring 12 hours each day to cut through a massive tree.

Herbert Christie was the boss of Falk's choppers for 50 years and one of the best fallers in the county. Herb would drive a series of stakes in the ground for a distance equal to the height of the redwood he was falling, often over 300 feet. When the tree was felled it almost always landed dead center on the stakes he had laid. It was not uncommon to see a group of city folks come out to visit the falling site, hoping to catch a glimpse of one of the big trees as it came thundering to the ground.

Before a redwood was toppled, a dozen or so smaller trees were first felled in its path to form a soft bed and cushion the tremendous impact as the tree hit the ground. To insure an accurate and safe landing, the highly-skilled fallers could make one of these trees hop, skip, or jump if necessary. When cutting down the big trees in steep terrain, the choppers always felled them up the hillsides. This allowed the tree to fall a much shorter distance, and eliminated some of the impact. If a faller missed his mark and broke up a tree, one could expect the air to turn blue with curses. Herb was probably one of the few exceptions to that rule. He was cool-headed, and on the rare occasion when his tree missed its landing, Herb knew that all the cursing in the world would not earn him a second chance at falling it.

After the choppers had fallen the trees in an area, "ringers" and "peelers" came in to cut off the limbs and remove the eight- to ten-inch thick bark. This waste material, called slash, was piled up, and during the dry month of September, fires were set to burn away the brush and tree refuse. The possibility existed that one of these fires would get away from the burn men, so the crew readied up to 200 five-gallon coal oil cans filled with water, and placed them in convenient locations. These cans had handles attached for easy maneuverability. But even with this precaution one of the big logs occasionally caught fire during the burn and was destroyed.

After the land had been cleared by fire, the "buckers" arrived with their long crosscut saws and cut the trees into movable lengths. Sometimes the two men on a bucking team did not see each other for half a day, since many of the trees they were sawing through were over ten feet in diameter and they were unable to see over the top of them. To perform their work, the men had to develop an acute sense of rhythm together. When the last tree was sectioned, the initial logging procedure was finished and the sectioned logs would wait until the following summer, when crews and equipment would return to remove them.

MOVING THE LOGS

It took a tremendous amount of labor just to set up the equipment needed for log removal. It was first necessary to build a railroad spur into the area, since the train was essential in transporting the machinery, tools and men to the site. A "landing" was then constructed at the end of the rail spur, usually at the base of a hillside, near a natural drainage, such as a creek bed. This natural drainage would later become the main logging chute.

Anchored to the landing itself was the "bull donkey," the center of the log removal operation. From its fixed location, this machine could pull 15 logs at a time from as far away as a mile. Essentially, the bull donkey was a steam powered "fishing pole" except that its "catch" was huge redwood logs.

In front of its steam boiler, the machine had three drum reels which operated two wire ropes, the "bull line" and the "haulback line." The bull line was used to pull the logs down the main chute to the landing. This wire rope, which measured over an inch in diameter,

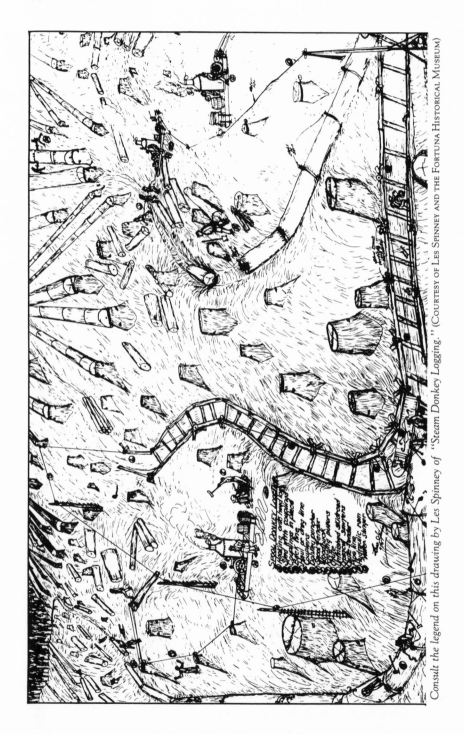

Consult the legend on this drawing by Les Spinney of "Steam Donkey Logging." (COURTESY OF LES SPINNEY AND THE FORTUNA HISTORICAL MUSEUM)

stretched more than a mile in length, weighed seven tons by itself and wound onto a single drum reel.

The other two drum reels operated the haulback line, which was almost three miles long and traveled in a continuous loop, like a clothesline, from the bull donkey to the top of the main chute, and back. This cable was ¾-inch in diameter and weighed 7½ tons. As it traveled up the hill, the haulback line was guided by spools until it entered the tail block, the other end of the "clothesline." The cable was then directed back down the hill toward the bull donkey through a series of smaller blocks which were suspended from trees and tall stumps. The haulback cable ran continuously, and was used to tow the bull line, as well as equipment sleds, to the top of the logging site.

Only two men were required to operate the bull donkey, an engineer, and a "woodbuck," who stoked the boiler with split wood. When this mechanical dinosaur was in operation, its huge boiler rumbled, and the large drum reels squeaked and groaned as they wound in the steel ropes. Once the bull donkey was in place on the landing, men and equipment headed into the woods for the final preparations.

The main "log chute" was cleared of brush and debris by the "swampers." Then, the steam donkeys, smaller cousins of the bull donkey, were winched up into the high areas of the logging site. Several of these machines were used in the logging site, and were stationed on the ridges which surrounded the main chute. Once they were anchored down, the 8-man donkey crew could reel in any log within reach of the machine's 800-foot cable.

It took three men to run the machine itself. Assisting the engineer and the woodbuck was the "spool tender," who coiled the cable as it was reeled in. The other five men were responsible for securing the steam donkey cable to the logs that would be moved. The "hook tender" was in charge of attaching blocks to the logs. He was assisted by the "second tender." These blocks had to be set just right, or a cable might snap loose and cut a man in half. Many lives were lost over the years from whipping cables. After these men had set the blocks on the tree to be moved, "rigging pullers" would thread the heavy wire rope through the blocks and attach the end of the cable, called the becket, to a nearby stump. Then the log was ready to move.

The logs were manuevered to the top of "pull chutes," at which point they were unharnessed and sent sliding down the hill toward the main chute below. Along these chutes a "waterslinger" wetted the track, which enabled the logs to slide freely. This man got his water from a series of small tubs stationed near these log runs. On a dry hillside, the waterslinger had to have water brought by horse to fill the tubs. The bull

Falk steam donkey crew. The invention of the steam donkey revolutionized redwood logging, replacing the earlier oxen teams. Note the large hooks used to anchor the machine down. (COURTESY OF CLARKE MEMORIAL MUSEUM)

line's water tubs, however, were always connected by a series of small flumes which were fed from a nearby spring or a hydraulic ram. If the pull chutes were too wet the logs could slide out of control, so the waterslinger also kept piles of dry earth nearby. He applied the dry soil to slippery areas and, by this method, slowed the logs' descent.

When the big redwood timbers slid to a stop at the main chute, the "coupler" then chained them together, end to end, and placed a thick piece of bark between each one to act as a bumper since the logs knocked and ground against each other all the way to the landing. The lead log (which axe men had beveled in the front for easier sliding) was always the largest in diameter. The rest of the 10 to 15 logs in tow were tapered in size, and the last one was perhaps two feet in diameter.

The man who was in charge of this string of logs from the top of the main chute to the landing was the "zoogler." He set the rigging on the lead log and attached it to the bull donkey's wire rope which pulled the load down the hill. Evan Rushing, Jr., was a zoogler in the Falk woods for ten years. He and his brother, Lloyd, took over the bull line in 1916 and together they ushered hundreds of "log strings" down the main chute. Although in title Lloyd was a waterslinger, the brothers actually traded jobs when they rode the logs down the hill—one of them keeping an eye

The "landing" of a woods operation. The landing was the end of the rail spur into the woods.
(Courtesy of Harry Falk III)

on the couplings and rigging, while the other slung water, and also made sure that the haulback line remained in its spools. If it was not, one of the men would have a long walk back up the hill to fix it.

When the logs reached the landing, engineer Jesse Barnes backed his locomotive and log-hauling cars into place. The landing was built to the same level as the cars, so loading the logs was an easy task. A series of pulleys were located the length of the landing; opposite them were the logs and a small steam winch. When the Rushings had recovered all of their rigging from the logs, the winch operator loaded the train cars, one at a time. At this point the cycle of early logging procedures was complete, except for the final detail which was performed by the zoogler.

It was Evan's job as zoogler to see that the couplings and head rigging were loaded onto a sled which was known as the "pig." This was simply a wide, hollowed log that could be attached to the haulback line, and was used to deliver men and equipment up the main log chute. While the pig was being loaded, the bull donkey's engineer also secured the main cable to the idle haulback line. When all was in place, and the Rushings were aboard the sled, the winches were engaged, sending men and wire rope back to the top of the hill.

During the return up the chute, Evan had to jump off and thread the bull line into the large spools that were to guide it down on the next log delivery. He also had to thread the haulback line if it jumped out of a spool. Evan had to be quick on his feet to accomplish this, considering that the 1⅛-inch diameter steel rope was in constant motion, and the sled did not stop. Sometimes Evan had to run as fast as he could just to catch the sled he had been riding.

The pig was a very handy tool for most of the chores it was called on to perform, including noontime food delivery. It had just one fault. It was prone to tip over.

At the top of the bull line a small cookhouse was constructed to provide the men hot lunches. The company found it easier to prepare the food at the main cookhouse and send it up to the logging site by train rather than bring all the men down from the woods to eat. Wooden boxes filled with meat, breads, vegetables, and pies were delivered from the main cookhouse to the landing, where they were transferred to the pig for the trip up to the bull line cookhouse.

Usually, things went smoothly while going up the haulback, but over the years the food sled managed to get upset several times. This always resulted in a terrible mess. When the sled turned over, a man suddenly found himself on his back in the wet and muddy log chute with upside down boxes, smashed pies, steaks in the mud, spilled milk, and soggy bread. He had to quickly leap up and locate two bare wires which ran along the bull line. By touching these wires together, a short circuit was created which rang a bell at the bull donkey and signaled the engineer to stop the winches. The person who shorted these wires many times received a low-voltage shock, but this was the only way to communicate with the engineer, who couldn't see the loads he was moving. After reloading the spilled boxes, the delivery man again shorted the wires together, signaling the engineer that all was clear to resume the journey toward the summit. Once the man reached the upper cookhouse, he then had to deal with the angry men who had to eat a mangled lunch.

While the zoogler and waterslinger returned up the main chute, other workers secured the loaded logs onto each rail car for the trip down the canyon. During this steep descent, the locomotive, with its string of cars leading the way, acted only as a brake for the tremendous load. The high wooden trestles they passed over creaked and swayed under the weight, while the slow-moving train gradually made its way down to the millpond.

RUNAWAY

One spring, a near accident showed the dangers involved in running the log train down the steep upper canyon. It was the beginning of another season in the woods and work was going slowly as cobwebs were cleared from minds and machinery. The crews were pulling logs at a new landing and the first train of the year was being loaded for the trip down the canyon. Jesse Barnes received word that his load was secured, and he released the engine's brakes. He was driving a new Heisler locomotive which had replaced the older Baldwin.

Walking up the tracks that morning from the logging camp was Jesse's nephew, Ted Barnes. Ted worked in the woods with a donkey crew, but had been in the logging camp that morning to take care of an errand. While crossing one of the taller trestles over the river, Ted heard the train's whistle from farther up the canyon. He walked a little faster to make sure he would be off the trestle when the train arrived, but the whistle was quickly getting louder. It was not the occasional toot he was used to hearing but almost a continuous blast which echoed down the canyon. Ted was nearly to the end of the trestle when the log train rounded the corner and he realized it was out of control. He leapt from the bridge, and fortunately had only a short distance to drop before landing on the ground. Ted looked up in time to see the log train thunder onto the trestle, its log cars reeling, and his Uncle Jesse in the cab of the Heisler as it pitched violently back and forth. The train barely stayed on the high trestle. It took only a moment for the roar and clattering of steel to pass by, and when it disappeared, Ted was left wondering if his uncle was going to survive the ride.

Apparently, when Jesse had started his descent into the canyon, the train lost its grip on the unused, rusty rails. The Heisler locomotive, being a geared type, could not be reversed while in forward motion. Several men had been aboard when the train left the landing, but they had jumped off when they realized what was happening. Jesse and his fireman, Erwin Foss, both decided to stay with the locomotive. They could not jump off and allow the runaway log train to speed unattended down the tracks toward town. To avoid that potential disaster, the two men chose an uncertain destination—either the log pond, or somewhere in the bottom of the river canyon.

When the train rumbled through woods camp, it was still out of control and rocking so wildly that it disintegrated the blacksmith's landing. But luckily, the section of track below the camp had been used quite frequently, so the rails provided better traction. Also, the incline was not as steep. When the engine reached this point they were able to regain control, and slow the load for the approach into town. Upon safely reaching the log pond, Jesse and Erwin both felt born again. They had been convinced that their runaway train had been bound for the promised land!

Sounds Of The Mill

FALK'S MILL WAS THE HEART OF THE WHOLE TOWN. IT was not the largest mill in the county, but it had sufficient modern machinery to cut 1 million board feet of lumber a month, making it a very successful mid-sized plant. The trees that were harvested and gathered in the company's forests continually filled the log train, which delivered them to the mill pond for storage and future processing. The mill's processing was conducted practically year-round, and every work-day the plant's machinery filled the air with sounds.

At five o'clock in the morning, the town came alive when the mill's steam whistle blew three long blasts to wake everyone, a bellowing that could be heard five miles down the valley. A half-hour later, two more whistle blasts signaled breakfast in the cookhouse. The whistle was not always the official breakfast call. When Jack and Minnie Randall cooked, the men did not listen for the whistle, they waited for Jack to play his saw. He pounded out his rhythms on an old circular sawblade that hung outside the kitchen. This sound was followed by the thumping of 100 pair of work boots as the men filed into the dining hall for breakfast. Most of the men who ate in the cookhouse were mill workers, because the woodsmen ate their meals at the upper camp, which had its own cookhouse and cook. Jack and Minnie were excellent cooks. Jack was also a good poker player, and it was a common sight to see two or three of the crew in the kitchen, helping to prepare the meal in order to pay off their gambling debts. At breakfast's end one final whistle signaled the start of another workday.

The first machine to operate was the plant's huge steam engine that powered the entire mill. It began with a whooshing sound that gradually

The Elk River Mill. Although it wasn't the largest mill in Humboldt County, it operated efficiently and shipped a considerable volume of lumber and wood products to all parts of the country. (COURTESY OF JIM AND VEDA WRIGLEY)

increased in volume. All day long two men with wide wooden shovels fed sawdust into the engine's boiler to maintain a constant steam pressure. Edward Newell was the company machinist, and along with operating the big steam engine he also maintained all the machinery throughout the mill. When the steam engine reached full power the plant was ready to saw logs.

While waiting for processing, the logs were stored out in the millpond area, where a crew of three men worked at positioning them. Jim McManus directed all the log handling around the mill and was assisted by his son, Albert, who operated the little Gypsy locomotive. The millpond crew had a variety of chores, which included unloading the log train, raising sunken logs, and stacking logs for milling. Of these three jobs, unloading the log train was without a doubt the easiest. A two-foot-in-diameter log was placed at an angle across the tracks and secured just above the height of the railcars—this was called the "jill poke." When the slow-moving railcar passed beneath it, the loaded log was sheared off the car and crashed onto a wooden ramp, where it rolled freely into the pond with a big splash. Some of these logs were so green, they immediately sank to the bottom. This created the term "sinkers." The pond was over 20 feet deep in places, and when a log sank to the bottom, its recovery was no easy task. This was especially true of the "butt logs" from the base of huge trees, which constituted most of the sinkers.

To retrieve a sunken log, a winch platform was secured between two high-floating logs, and a cable was lowered into the murky waters. The millpond worker blindly tried to snag the submerged tree section with a large hook that was on the end of the cable. When the sinker was snagged it was winched up by hand and secured to the two floats. Using the same procedure, the man then raised the other end of the log and moved it to a spot where it could be pulled from the pond. The log pond itself was not intended to be used as a storage area, but more as a convenient holding pen until the logs could be moved to their intended destination, which was either the "cold deck" or the mill. The cold deck was the log storage area, but during the summer months when logs were being removed from the woods, more of them were stockpiled on the deck than were sent through the mill. This created a surplus of logs that enabled the mill to produce lumber year-round.

The industrious little Gypsy locomotive was used to remove the logs from the pond and place them on the cold deck. With its high-pitched whistle, its leaky boiler emitting steam noises, and its bell ringing amid the clatter as it jumped up and down on the tracks, the Gypsy sounded like it was ready to explode. But anyone who was familiar with the little veteran knew otherwise.

Ted Wrigley knew the Gypsy well after working several summers around the millpond with Jim and Al McManus. He never forgot his first day on the job when he mistakenly opened a rail switch backward and brought the Gypsy to a metal-grinding halt as it was thrown off the tracks. Al soon had the machine back on the rails with the aid of a couple of tree jacks, and the only damage that occurred was to Ted's pride. The machine had been through a lot of abuse over the years and one derailing was not enough to end its career.

Maneuvering the big logs from the pond to the cold deck presented no problem for the Gypsy, but when it came to stacking the logs in piles, the machine was of little use. Some of the cold decks were over 30 feet high, and it was the crew's job to stack these logs by hand. This extremely hazardous job was performed by a man on the deck who used a series of geared jacks to slowly raise a log, one end at a time. These timbers weighed tons, and the worker was in the precarious position of being under the heavy log while he carefully cranked the small jacks that supported it.

One day, Jim McManus was jacking up a big log near the top of the cold deck while Albert was engineering the Gypsy, which had a safety cable stretched from its spool to the log. The timber was supposedly secured and Jim pounded several holding wedges beneath it, then

The bandsaw replaced the earlier circular saw, which was often too small in diameter to cut through big redwood logs. Noah Falk was one of the first people in the redwood timber industry to see the potential of this new type of saw. This picture was taken at the Scotia Mill in 1900. (COURTESY OF STAN PARKER, PACIFIC LUMBER COMPANY)

signaled Albert to back off the cable. When the tension was released all of the wedges snapped—the huge redwood rolled from its position and pinned Jim beneath it. Albert hit the winch with all the power the engine had. The enormous log lifted, but only momentarily, because the cable slipped from the Gypsy's spool. The log rolled back onto Jim, and down the deck.

In a matter of minutes a special train was prepared and the tracks of the Bucksport and Elk River Railroad were cleared. The severely injured pond boss was placed aboard a flatcar and Albert and a friend were by his side. With its bell ringing, the unscheduled train made its way down the valley and people along the way realized that another accident had occurred. The train passed directly in front of the McManus home. When Jim's youngest daughter, Lelah, recognized that it was her father who lay injured on the flatcar it was all she could do to hold back her tears. He waved to her as they passed but he never returned home. Jim died three days later in the hospital. The accident was especially hard on his son, Albert, who had tried so hard to save his father's life.

In all phases of the timber industry danger was close at hand, and for even a veteran with over 20 years' experience, a sudden miscalcula-

The edger saw in the Elk River Mill. Noah Falk is pictured on the left, closest to the camera. (COURTESY OF JUDGE HARRY FALK)

tion could produce an irreversible tragedy. Working amidst timbers that weighed tons, high-tension steel cables, and whirling razor-sharp saws, the workers knowingly engaged themselves in the dangers of "Redwood Roulette." But no matter what their personal fate, the logs kept moving.

INTO THE MILL

Each day, the first log to be sawn was floated into a small channel that led from the pond to the mill. A continuous-running chain with teeth pulled the log from the channel, and moved it to a staging area where the log sat until the bandsaw carriage was in position. The log was moved onto the carriage by a steam-powered piston sledge. When the sledge punched the log onto the carriage, the mill, built as it was on a foundation of sawdust, shook as if it had been hit by an earthquake.

The log had to be placed squarely on the slip carriage, and moved around several times during the milling procedure. All of these adjustments were done by hand and were the job of the "carriage setter." Since this was the most physically demanding job in the mill, the carriage setter was usually a very brawny individual. He used a rachet setup to move the log around, but because each log weighed well over a thousand pounds, his job was of great difficulty.

The director of the milling process was the sawyer. The Elk River Mill employed Harrison Fryers and Ed Callihan for this critical job. A good sawyer required a shrewd eye and a calculating mind; otherwise, much wood might be wasted through faulty cuts. These men chose their cuts according to the size of the lumber needed and the grain structure of the log. They could estimate at a glance how many board feet of lumber an entire log held even before it was milled.

The bandsaw created a loud hum that was audible only around the mill, but when the slip carriage passed a log through its blade a piercing howl was heard throughout the town. After the log had passed, the sound diminished to the hum of the blade.

Keeping the bandsaw blades sharp was a full-time job for two men. They needed plenty of light for the tedious process of filing the hundreds of teeth on each saw blade, so their large shop area was surrounded by glass windows on three sides. These "filers" worked in an endless round because the bandsaw's blade needed sharpening almost every day. They were required to have two replacement blades ready for use at all times, while two more were being serviced in the shop. Along with this responsibility, the filers also maintained scores of circular saw blades, planers, and other cutting tools that were employed throughout the mill.

When a timber was passed through the bandsaw it was transformed into planks of wood which sometimes measured over six feet in width. These sheets of redwood were placed on roller-topped tables, and moved first to the edger and then to the trim saw. Both of these were circular saws.

The edger, like the bandsaw, ripped the wood along its grain, and made the wide planks into standard-width lumber. The trim saw, on the other hand, was a crosscut tool, and it was used to cut the boards into various lengths. The intermittent sounds of these saws were barely heard in the town, but in the mill they were highly audible. During each passage of lumber, the edger produced a loud ring, and the trim saw punctuated the air with its sudden, short, shrill bursts.

After the sawing, the processed lumber was again sent down roller-topped tables where it was sorted and graded. Most of the rough-sawn wood was stacked on hand trucks and moved to the mill's shipping platform. The remaining lumber, usually the finer-grained pieces, was sent to the resaw room to be processed into finished products.

In addition to lumber, the Elk River Mill also produced shingles and pickets from the rougher grades of redwood. The scream of the shingle planer was the loudest sound in the town as it chewed through the tough-grained woods. The only remaining sound was that of the chain elevator. This metal contraption squeaked and rattled all day long as it carried wood scraps from the mill to the burnpile outside.

The bizarre chorus of mill machinery came to an end at six o'clock in the evening when the final whistle blew. The crews went to their homes and cabins to clean up for dinner, and the town, once again, settled into the quiet of the valley. At night, the only sounds were of crickets, an occasional voice, a barking dog, a car thumping across the wooden plank road and, if one listened closely, the flowing waters of the Elk River.

Town Life

THE TOWNSPEOPLE OF FALK ENJOYED A REAL FEELING OF
camaraderie, which grew out of their work. When they had time to
socialize, one of their favorite spots was the General Store. The name
"General Store" was a most appropriate description for the only building
in town that had a cash register. For many years Matt Carter managed
the company-owned store, and over his counter he offered food supplies,
hardware items, mail service, and everything from lanterns to shoes. But
the store also offered people a place for informal gathering. During the
warm summer months, Matt reopened the store in the evening after
supper, and people stopped by to visit, play cards, or buy soft drinks. The
youngsters lined up at the counter to spend their pennies, and each
evening became an impromptu social call for all ages.

Late afternoon gatherings around the store were of a slightly different nature. This was when "the guys" got together. At the end of a
workday men sat on the steps outside the store and engaged in friendly
contests or swapped stories. Since many of the workers had migrated
great distances before settling in Falk, countless tales were told, some
with foreign accents. The men rolled dice, played poker, or tried their
accuracy at knife throwing. The game they played was called "splits."
The first player threw his knife into the ground and the second player
attempted to stick his knife as close as possible to the first.

The men who had energy to burn after their day's labor often
enjoyed a game of basketball, which was made all the more interesting
because the players still wore their heavy, caulked boots. For some of the
more hardnosed characters who roamed the woods and mill areas,

however, these afternoon get-togethers were not always to exchange pleasantries. Hanging in the general store were two pairs of boxing gloves. When disagreeing parties could not find common ground to resolve their differences, they laced on the gloves. It was a healthier way to resolve problems than broken bottles and bar stools, and some of the men at Falk were very handy with their fists. These gloves settled the majority of quarrels, and for the most part, boxing and wrestling developed a friendly competition among the men. The front office was surprisingly well represented in these matches. Stories say that two of the best dukes in town were Lloyd Rushing, who worked on the bull line in the woods, and company officer Winfield Wrigley.

Apart from these outside activities and the services over the counter, the store was also an excellent place for a young person to receive job experience. The part-time work amounted to only a few hours each day, which was a marked contrast to the 12-hour workdays required by most other company jobs. During his high school years, Leland Newell was employed by Matt to deliver groceries and to sweep out the store. Most of the families ordered their staples in bulk, and Leland had to muscle the 100-pound sacks of flour and grains from the storeroom onto the flat bed of the old wooden-wheeled delivery truck. He then drove along the town's corduroy road—so-named because it was paved with wooden planks—to make his deliveries. One summer when there was a decline in orders, Matt told Leland that he could take the truck and go camping wherever he wanted. Cars were a relatively new phenomena at that time and only a few people in the valley had them, so Lee was thrilled. He and a friend packed their camping gear and drove to a remote area several miles away. For three days, the two vacationers fished and sat around their camp, enjoying the good life. Even in a little town like Falk, it was nice to get away for a while.

To keep food purchases down to a minimum, many people maintained gardens, a few chickens, fruit trees, and whatever else they could manage. For the ones who preferred to buy groceries, the general store sold everything from bread and milk to meat and vegetables. The families shopped during the day and, if the order was not too big, they carried it home themselves. Usually some energetic youngster was nearby who would help his mother with groceries, but sometimes these children had a little too much energy, as in the case of nine-year-old Leslie Steenfott, who had been quite a character in Falk before the family moved to Eureka.

Leslie was a live wire, and once he was in the store he knew no bounds. When his mother was not looking he was up and down the aisles investigating everything. Mom tried putting him in a corner, but Leslie

Falk viewed from a hillside above the mill. Center left is a cluster of small buildings including the gas station, the freight station, the general store and company offices. Crew cabins and homes are scattered down the valley and on the hillsides. During the community's heyday, over 400 people lived in or nearby the little lumber town. (Courtesy of Evelyn Oberdorf)

could find some sort of mischief wherever he was. Matt kept a keen watch on young Leslie, and one day he devised a way of containing him. When Leslie and his mother arrived, Matt met them on the porch of the store where a freight boom was located. He swung the boom hook over the porch and attached it to the back of Leslie's coveralls, then swung him back out so that he hung several feet off the ground. There he kicked and flailed without upsetting anything. When Ethel was done with her shopping, Leslie was taken off the hook and loaded with groceries for the walk home. This method of containment proved to be quite successful, and at the same time created a very humorous sight for passers-by.

Along with its over-the-counter service the general store supplied the food for the two cookhouses. For a number of years the company had tried to raise its own cattle and pigs for slaughter, but it was too much trouble, so they returned to ordering meat through the store. Matt bought his meat supplies from either the weekly meat wagon that came from Eureka, or from George Wrigley. George had regained his health after his food poisoning, and with the help of his son Irving, raised beef and sold it fully prepared for 12 cents a pound.

After Red Braghetta left the general store to work in the company office, Matt hired a new assistant storekeeper, Oliver H. Blaikie. Oliver and his young family had come from New Brunswick and, like others before them, had ended their journey at Falk. Since many of the residents who lived in the Upper Valley were also from the Maritime Provinces of Canada or the New England area, the Blaikies found much in common with their new friends. After his first year in the valley, Oliver had grown to genuinely love the town and its people and was described as "a man with a heart of gold" because he never refused a person who needed help. He had little money and what he did earn at the general store was barely enough to support his family. Despite this fact, Oliver still assisted many of his friends in his free time. It was during the Blaikies' third winter in town when misfortune struck the family. Oliver came down with tuberculosis which forced him to resign from the store until his health was restored. Since sick pay and disability insurance did not exist, the Blaikies were without income and almost penniless. Several of Oliver's friends came to the general store to discuss the situation with Matt. They told Matt that they would take turns doing Oliver's storework for him if Matt would continue to pay his salary to the family until Oliver returned. This was agreeable to Matt, who also recognized the poor man's plight, and wanted to help. True to their word, Oliver's friends took turns working at the store during the next weeks and his income was provided.

This willingness to help someone in trouble reflected many of the townspeople's feelings toward each other. Vernon Olson remembered those feelings:

> "Folks really looked out for each other back then. If someone got sick or disabled for a while, then friends and fellow workers gathered up donations for the person. People stopped by, brought you food, put your garden in for you, chopped some wood—just did whatever they could to help you out. Those were happy days, living in Falk."

Life at Elk River was a grass roots existence. If one did not make a living in the woods or mill, ranching or farming were about the only other alternatives. If a person desired a higher paying job, it was necessary to move elsewhere. But some of the people of the town tried to add a bit of culture to their backwoods pleasures by forming the Falk Literary Society. Their Friday evening shows were events that whole families enjoyed, and out of the 400 people who lived in the Upper Valley, quite a few were willing to entertain their friends in one capacity or another. Acting out humorous plays was popular, along with playing musical instruments, singing songs, or reading poetry and verse. On special occasions, the Literary Society would hire a vaudeville act from

The town of Falk. The Hotel-Cookhouse is to the left and Jonas Falk's two-story home is visible in the upper right hand corner. The firewood shed is in the center of the photo, and to its right is the company garage. (Courtesy of Evelyn Oberdorf)

out of the area, but it seemed that no matter what the entertainment was, the audiences were always wholeheartedly appreciative.

Though the town of Falk was located only eight miles from Eureka there was a sense of remoteness about it. It was created, in part, by the geography of steep surrounding hillsides, but mostly by the interdependence fostered among the people. The community encouraged local social gatherings in which everyone could participate, and on some occasions entertainment was brought in from other areas. Store-bought luxury items became attainable through a number of sources, but if one desired certain items immediately, it was necessary to make a trip to Eureka.

For those who had to make that trip, public transportation was offered by the Nellist Brothers Stage Lines which ran twice a day between Falk and Eureka. People could catch the stage at the general store, or have their goods delivered to their home for a modest sum. Oscar Nellist, who had managed Falk's general store for several years before starting the stage line with his brother Frank, was highly accommodating to the needs of the valley residents. He delivered such items as catnip, spirits, repaired shoes and watches, hardware, medicines, and even pets (they rode for half fare). Since no public school bus system existed, the county subsidized the fares of high school students who also rode the stage, paying $5 of the $8 monthly fare. If the five-seater Pierce Arrows became too crowded, the students had to make room for the extra riders. Occasionally, the stage could be seen bouncing along the valley's gravel road heading for town, all five seats filled with people, a dog on someone's lap, parcels strapped to the roof, and several youngsters riding the fenders and running boards.

Fine goods and crafts also became available from traveling peddlers. The peddlers traveled in large horse-drawn wagons which were filled with bolts of dress materials, colored yarns, shoes, perfumes, jewelry, and countless other items that tempted the rural shopper. These wagon entrepreneurs would take furs and other goods in exchange for their products. Some traveling salesmen offered entertainment shows. It was through them that many of the townspeople had their first chance to hear recorded music or view a silent motion picture.

One of the local woodsmen, Les Christie, was so thrilled with his experience of hearing recorded music that he sent away through a mailorder catalogue for a phonograph which played wax cylinder recordings and mechanically projected its sound through a large horn. Les had saved for several months to afford the machine and when it arrived, he almost wore out his only two records while demonstrating its sound to curious listeners. Some of these people thought Les's player was the funniest thing they had ever heard.

Traveling novelty acts were not the town's only source of entertainment. The residents of Falk were also quite resourceful in creating their own festivities, which ranged from community picnics and talent shows to the more lively Saturday night dances. Because people had so much work to do, most of their social gatherings occurred in the evening. The annual Fourth of July picnic celebration was an exception.

Early that morning, the lumber train idled into town and parked near the general store to await preparations. Benches were secured to the flatcar decks, and colored streamers were draped from one end of the train to the other. When the job was finished, the engineer gave a few blasts on the steam whistle to signal the townspeople that the annual picnic was about to commence. Families arrived at the loading platform carrying big baskets of food and dressed in their best Sunday clothes, all eagerly anticipating the day's events. The train crept slowly down the valley toward the Jones Prairie picnic grounds, sounding its bell and whistle at each of the many stops to pick up the people who waited along the tracks outside of town.

The picnic grounds were a superb summertime gathering spot and had plenty of room for the youngsters to run around, an outdoor hardwood dance floor, tables, lawns, big shade trees, and the nearby Elk River. After a few hours of games and socializing an enormous feast of homecooked foods was laid out on the tables and everyone shared in the enjoyment of a communal meal. After eating, there were more games and socializing. This was a time for everyone to forget their work and have a day of leisure with their family and friends.

Hard
At Play

IN A LUMBER TOWN WHERE CREWS WORKED SIX DAYS A
week, Saturday was the only night a party was allowed to carry on into
the early hours. The Saturday night dances were the most gala of events
in Falk. People came from Eureka and other outlying areas to attend
these affairs. So popular were the dances that Falk quickly outgrew its
original dance hall and a new hall was built near the Elk River Dam.
Among the many bands that were hired to play was Barkdull's
Orchestra, which played around the county from 1917 to 1924.

Eli and Charlotte Barkdull were brother and sister, and the leading
players in their band.* They traveled to lumber camps in the hills from
Weott to Trinidad where they performed for packed halls. Charlotte
always remembered the nights they played at Falk because the
enthusiastic crowds would not let them leave. They were hired to play
from 9:00 p.m. to 1:00 a.m., but it always seemed that by 1:00 a.m., the
good time was just beginning. When the last tune was announced, the
cheering, stomping crowd would pass hats around the room to gather
more money for the band. This process went on until three or four in the
morning, at which time everyone was finally too tired to continue.

Along with music and dancing, the Saturday night social included
a midnight supper. When the clock struck twelve the band stopped,
tables were set up, and the women brought out fried chicken, potato

---

* The members of the band were: Eli Barkdull, Saxophone; Charlotte Barkdull, Piano;
Arleigh Noah, Violin; Mansel Clark, Coronet; Arthur Remell, Clarinet; Kenneth Hill,
Trombone; Austin Corbett and E. Sundquist, Drums.

Charlotte Barkdull of Barkdull's Orchestra performed at the Saturday night dances in Falk. She later married Frank Niskey, a woodsman from Falk who studied law at night by kerosene lantern and went on to serve over 20 years as a county judge. (COURTESY OF CHARLOTTE NISKEY)

salad, cakes and freshly-baked pies. Sometimes a lady filled a beautifully decorated basket with home baked and canned goods and auctioned it to the man with the highest bid. That man became her date for the rest of the evening. Following the supper and social events the tables were removed and the music was resumed.

While most youngsters did not attend the dances, there was one young man who saw them as an opportunity to make some money. When Chester Barnes was 10 years old he operated the dance hall concession stand. He sold flavored soda water to the dancers, but there was not much profit in that. The big money was in lemonade. Chester started his preparations early in the morning, buying cases of soda water from Matt Carter's general store. To protect his investment from all the thirsty kids during the day, he loaded the soda aboard a railroad handcar and pumped it down the tracks to a secret hiding place where he concealed the bottles to keep them cool. When evening arrived, Chester mixed the lemonade and then retrieved the soda from its hiding place. Although public drinking was not tolerated at these dances, some people brought whisky bottles with them. They gave their bottles to Chester to keep behind his concession stand, and he poured them a drink at their request. There were always a few bottles of "Yellowstone" or "Myrtledale" behind the counter, and their owners developed secret passwords when they asked for refills. One man would come up and say, "I want the key to Myrtledale." Chester then set up his drink, with a chaser of lemonade, and no one knew the difference. For this under-the-counter service he received extra tips. Sometimes he made over $20 in a night—about the same wage a foreman in the woods made for a week's labor.

These social gatherings, all of which were free, became an integral part of community life in the Upper Valley. For many, the anticipation and memory of such events helped to carry them through a hard week of work, whether it was at home, in the mill, or in the woods.

FISHIN' AND DRINKIN'

With a 72-hour work week, Sunday became the only day people were free to enjoy personal pastimes. These welcome breaks often meant either sleeping in late after a Saturday night celebration, or spending a quiet afternoon fishing by the riverside.

On Saturday nights the dam was opened to let the trapped mud out of the millpond, and it was not closed until Sunday afternoon. While the dam was refilling, the water level at the base of the spillway dropped substantially, and many large salmon and steelhead became stranded in the deep pool. Although some of the fishing techniques would make a game warden cringe, people of all ages tried their hand at catching a few of the trapped fish.

As a young boy, Leland Newell often went beneath the dam when the salmon were spawning and splashed around until he scared a 50-pounder out of the water and onto the land where he could grab it. If he did not want to go to all that trouble he simply put a gate in the fish ladder and speared the fish when it had to stop. When Leland got a little older he began to feel guilty about his techniques, and eventually took up the traditional hook and line method.

In the early days the Elk River was filled with so many schools of salmon and steelhead that even local dogs were able to land them from the river's edge. But these large fish were only present during the spawning season. The trout, which was the most abundant fish, could be caught year-round. Because fish were so plentiful, a fisherman did not have to be highly skilled, and the limit was often whatever he could carry home with him. The record for the most fish caught at one time in the Elk River can probably be credited to Chester Barnes when he was a youngster. His mother, Loleta, had taken their horse and buggy into Eureka early one morning, and Chester planned to catch a few trout for the family's supper. When Loleta returned that afternoon she found Chester in the kitchen amidst a huge pile of fish heads, cleaning his 112 trout!

Along with the many fish that inhabited the Elk River there lived another water dweller, the eel. This unusual aquatic creature was not desired for the dinner table and was generally overlooked, but it managed to make its presence known in mischievous ways. One winter, the water level in the log pond kept dropping, yet no visible leaks existed in the dam itself. This went on for several months until the company decided to investigate the problem and drain the entire log pond before the winter rains started. The dam was inspected and proved to be watertight, but it was discovered that a large group of eels had been stranded in the pond, and over the years had dug several escape tunnels, each over 20 feet long, under the dam's footing. The tunnels were plugged and large amounts of gravel poured at the dam's base to prevent any further erosion.

This problem with the eels turned out to be minor in contrast to the damage that ensued during huge winter rainfalls in the early 1920s, when a lengthy downpour swelled the river and sent a flood of timbers through the canyon, destroying several trestles and finally rupturing the dam. The flood-damaged properties were quickly rebuilt. The new dam was built exactly like the original one, and the deep pool at its base continued to be the best fishing spot in the valley.

Fishing was fun and even practical, but for some of the loggers it did not take the edge off a week's work as would a good pint of whisky. No liquor was available in Falk unless you counted a stomach reliever called "Tanalac" which was 16% alcohol and was sold in the general store. Although some of the men drank the stomach potion when nothing else was available, several enterprising men in town produced some very tasteful home brew. When Ed Danielson was not working in the woods he was the town constable, which was a good cover for an underground brewer. Ed's problem was that his brewing was no secret. Fred Johnson, the timekeeper, who had the keys to all company buildings, occasionally stopped by the building where Ed kept his brew and indulged in free samplings with his friends.

Some of the moonshiners, though, were not at all friendly if one tampered with their batch. Peter Oberdorf, one of the logging camp's blacksmiths, walked into the camp cookhouse one day, poured himself a cup of coffee, and picked up a donut. He was familiar to everyone who worked in the cookhouse because he occasionally helped them out when he was not too busy with his own work. A new cook had started that month and Pete decided to get acquainted. He stepped up to the stove where several big pots were bubbling away, stuck his nose over one, and said how good it smelled. The big, burly cook stomped over, grabbed Pete by the collar, and told him to keep the hell out of the kitchen. Pete did not stand around to ask questions, but he felt puzzled by the irate cook. About a month later, during which time Pete had avoided coffee and donuts, a friend came by his blacksmith shop and gave him a small sample of whisky. "This is great stuff," his friend said. "The cook in there makes it." Suddenly the kitchen episode made sense to Pete. He later got to know the cook but made a point to never sniff around his pots again.

A short time after Pete's moonshine experience, a hard liquor still was discovered in the hills above town and caused the local home brewers to lay low. Apparently Hiram Thrapp had been making more than railroad ties at his secluded work area. Moonshiners were not uncommon in Humboldt County and over the years many stills were discovered tucked away in the hills. The area had a large clientele of drinkers who preferred to buy the cheaper home brews. But for the most part, Saturday night in Eureka was the scene of the most intensive drinking.

Bill Furlong, a woodsman from Falk, made those Saturday nights a weekly event, and always returned back to his cabin late on Sunday, still tipsy from his drinking. One weekend, two of Bill's neighbors, Evan Rushing and his brother, Lloyd, decided to play a joke on him. After Bill had left for Eureka as usual, Ev and Lloyd borrowed a tree jack and jacked up one corner of Bill's cabin about a foot and a half. The brothers put a two-by-four under the cabin to support the structure, then pulled the jack out, and waited for Bill to come home on Sunday night.

Sure enough, Bill arrived home late, drunk again, and flopped down on his bed. Ev and Lloyd gave him a little time to get relaxed, then crept up to the cabin and knocked away the already straining two-by-four. The whole cabin came down with a crash, upsetting most of its contents, including Bill. The Rushings ran to their cabin and lay on their bunks as though nothing had happened. Meanwhile, Bill bolted out the door and ran directly down to Evan's and Lloyd's yelling, "Did you feel the earthquake? Hell of a big one. Nearly flattened the place."

"No," they said. "We didn't notice anything; are you feeling all right?" Bill didn't know what to think. He just kept muttering about the biggest damn earthquake he'd ever felt, and headed back to his cabin to

Russel Weatherby kicking up his heels in the family car. Those who owned an auto found it easy to go into Eureka for a taste of "big city" excitement. For those without cars there was a stage coach line, horse and buggy—or an eight-mile walk. (COURTESY OF RUSSEL WEATHERBY)

sleep it off. The following morning at work, amid the skepticism of his fellow workers, Bill enthusiastically related the story of the big quake that had upset his cabin and almost leveled the town.

MAN'S BEST FRIEND

One of the most colorful personalities in the town of Falk was an eccentric pit bull known as Old Man. Old Man shared a cabin with Evan and Lloyd Rushing and accompanied the two brothers in the woods. Their cabin was small, and had only two bunks. But Old Man was happy to curl up in the corner by the woodstove.

The dog had a unique fascination for riding on anything that moved, and during the course of the day he attempted to satisfy this obsession by every means possible. Each workday, the three roommates rose to the sound of the 5:00 a.m. whistle and made their way to the locomotive barn to catch the morning crew train up the canyon. The Rushings rode in the boxcar with the rest of the woodsmen but Old Man rode on the fireman's seat in the locomotive. If someone happened to be sitting there they were forced to contend with an adamant dog that insisted on sharing their seat. Within a few minutes Old Man had usually crowded the person from their spot. When the train pulled out, the canine hung his head out the window and panted like the steam engine, excited to be aboard.

Old Man's first stop of the day was for breakfast at the cookhouse. Here he ate his fill of table scraps in preparation for the all-day journey that lay ahead of him. After the morning meal, Old Man re-boarded the locomotive on his way to the logging site. On arriving at the landing, he jumped off the train and ran to take his place in the first sled that was going up the log chute. Sometimes, as a joke, one of the men would attach the sled's 20-foot tow chain to the fast-moving haulback. This always surprised the dog and sent him tumbling backwards off the sled. But Old Man was quick to regain his feet, chasing the sled until he could leap back on it.

Woods boss Jim Copeland was one of the dog's best friends. During the morning they toured the logging site together, and when Jim took his noon-time siesta beneath a tree, his four-legged partner stood watch to warn away any intruders. If someone wanted to locate Jim, they simply looked for Old Man.

Shortly after lunch, the dog went to the top of the log chute and waited for Ev and Lloyd to hook up the next set of logs that were to be pulled down the hill. When Ev had secured the rigging, Old Man climbed with him onto the lead log and stood at the foremost point, looking like a hood ornament. He remained there until they reached the log landing. Although by now it was well into the afternoon, the wandering dog still had miles to travel.

The dog seemed to know all the train schedules. When Jesse Barnes blew the whistle to signal that another load of logs was on its way down the canyon, sure enough, sitting next to him was Old Man. Upon reaching the millpond he would leap out of the cab and run to the mill's loading dock. If all went well, he made his last connection of the day, the afternoon lumber train to Bucksport. The firemen on the Bucksport Line were well aware that their passenger with the cropped ears demanded a window seat, and when the train rolled down the valley toward Humboldt Bay it appeared that a pit bull was operating the locomotive.

After arriving at the Elk River Mill's shipping dock, Old Man had two ways to go: either follow the lumber aboard ship and become a sea traveler, or catch the evening train back to Falk. The canine wayfarer always came back home, usually arriving around six o'clock in the evening, just in time for dinner. This was a day in the life of Old Man.

The Townswomen

THE LOGGING AND MILL WORK WAS A MAN'S WORLD, and for that reason little opportunity existed in Falk for women to acquire a profession. The only jobs available for women in the town were kitchen work in one of the cookhouses, or, if they were qualified, teaching at the Jones Prairie School. For most women, the choice was between moving to the city to learn a skill, as many of the young women did, or staying in the town to fulfill a more traditional role and raise a family. Those women who made the latter choice were vital to the community's success. The work that was involved in maintaining a household rivaled that of the woodsman, and the reward was a homespun existence, from the clothes the family wore to the food they ate.

At a time when kitchen conveniences consisted of a wood cookstove and a sink, food preparation was a time-consuming chore. It meant maintaining a large garden, picking berries, baking bread, cooking with whole grains, stocking a root cellar, and taking on huge annual canning projects that put fruits, vegetables, fish and meat on the storage shelves for the coming year. Some of these families had six or seven people in the household, so a substantial amount of food needed to be processed. Of course, a family of that size also produced many extra hands to help with the chores.

Along with the food preparation there was also a mountain of dirty clothes, all of which were washed by hand, and a continual need for mending and sewing which, if one were fortunate, could be stitched on a treadle sewing machine. Also at hand were a multitude of other tasks; making soap and candles, chopping firewood, taking care of the chickens

Traditional and non-traditional roles of women. Left, Ruby Fryers with twins Stedman and Stillman, first twins born in the Upper Valley. Most women had few career choices in a town like Falk. If they didn't wish to get married and raise a family, they had to leave town and seek their fortune elsewhere. Grace Rushing, right, grew up in Falk, but left to pursue a career as an opera singer named "Madame Henkell." (LEFT, COURTESY OF KEITH FRYERS; RIGHT, EVAN RUSHING)

and maybe a cow or a pig. The work at home was a continuous process, just as the logging was for the woodsmen, and like the woodsmen, the women had a certain peril to face—childbirth.

Living in a town that lacked a resident doctor created a great inconvenience for those that needed immediate attention. Realizing this, Dr. Curtis Falk, Elijah's son, tried to offer his services to the people in the town. With only a moment's notice, Curtis often rode out to Falk in his horse and buggy to tend an illness or injury which had occurred. Some of these were emergencies, however, and could not wait for a doctor to be summoned to the town.

On one occasion a pregnant woman, Mrs. Ida Fleckenstein, rushed into the company office and shocked Vernon Olson when she exclaimed that she was about to have her baby. Vernon knew a little first aid, but he was not about to volunteer his assistance in childbirth. He commandeered the only vehicle in town, and setting the world's speed record over a corduroy road, raced Ida toward Eureka. Vernon became nearly as desperate as his laboring passenger, but managed to reach the hospital just as the birth was about to occur. This obviously was not the standard procedure for every birth in the community, but it was an example of the vulnerability a woman faced at her crucial hour.

Most of the women who lived in Falk were content with rural life and the rewards it offered. But one young lady from Falk, Grace Rushing, followed her heart's desire to become an opera singer. Her path to recognition was a cinderella story, except that she possessed a silver-lined voice in place of a glass slipper.

The daughter of Evan and Louisa Rushing, Grace grew up in Falk, where her father worked as a zoogler. She shared the family household with three sisters and three brothers, and though they were quite poor, their father made sure they had a comfortable home, food to eat, and clothes to wear. At 18 years of age, Grace realized her future was not in the Elk River Valley, and she moved to San Francisco to take a job. In her free time she began to train her voice for opera and studied under several accomplished instructors. While living in San Francisco, Grace was married, and she adopted her married name on stage, calling herself "Madame Henkell." She spent several years in New York City, performing for large audiences, and toured the country, before settling down in San Francisco, where she sang in the City Opera. But Grace never forgot her roots in Humboldt County and returned for a public appearance at the Rialto Theater in downtown Eureka.

The path that led Grace Rushing out of the Elk River Valley was not a path every woman could, or would want to, travel, because it meant leaving behind family and friends, and all the benefits of a close community. On the other hand, an independent-minded woman who wanted a job outside the home found it difficult to live in such a small town. There was one exception to this general rule, Maggie Biord, who ran the logging camp cookhouse. Maggie became one of the only women to develop a successful career in the company town, bridging the gap between the role of the working man and the role of the working woman.

Maggie

Maggie's personality was colorful and outspoken, and she had an intense desire to offer the best food service possible. Her vitality was unique, and her energetic enthusiasm created a spirited life around the camp. Part of this spirit was a volatile temper. No one dared cross Maggie. Even the company bosses took their hats off when they entered the upper cookhouse. In fact, Falk's main logging camp was known as "Camp Maggie."

She was a married woman who really did not have to work, but she loved to cook. Although she left the camp several times, she always returned to the job, and over the years, built up quite a reputation in the town. Physically, she was a big woman, weighing over 200 pounds and endowed with a loud, booming voice. Maggie was a hard worker and expected everyone around her to be of the same fortitude when it came to getting the job done. Her kitchen crew prepared three meals a day for 100 men with robust appetites, so the preparations and cleanup were immense and continuous.

Maggie knew the type of meals men liked, and served them to the workers in huge portions. Her philosophy was that if a man left the table while there was still food on it, then he knew he'd had enough. On the breakfast table every morning were steaks, eggs, potatoes, donuts, pancakes, oatmeal, and coffee. The two 60-foot tables used for dining sagged visibly when loaded with the enormous weight of food and dishes. In a matter of 20 minutes, after "seconds" and "thirds," the dining hall emptied and the kitchen crew once again began the cleanup.

Her fiery reputation traveled further than the town, and sometimes it was difficult to locate people that were not afraid to work for her. She had very definite ideas about how a cookhouse should be run. Bus carts were not used to move the dishes and food because she felt they made people lazy; instead, everything was carried by hand. The plates weighed two pounds each, and the cups were thick, heavy and round, with no handles, so it took a lot of hustle to move what amounted to nearly three quarters of a ton of table settings per day, not including the food!

Thursday evening was the big dinner feed of the week. Maggie served steaks, chicken, fresh fruits and vegetables, two or three different kinds of pies, several assorted cakes—everything a hungry working man might dream about. She made all of her own bread and pastry, and sometimes cooked 40 or 50 pies at one time. The cookhouse received four sides of beef each week, and along with her normal cooking duties, Maggie completely trimmed and cut all of this meat herself. If any of the food was not top quality, it was returned or discarded.

Always in good humor, Maggie maintained a friendly quarrel with Matt Carter, the general store manager. He provided the supplies for the upper camp, and at one point he thought Maggie was serving too much food and cut back on her order. When the Gypsy arrived at the upper camp with fewer provisions than she had requested, Maggie laid into the superintendent who had delivered the order and told him that she was "gonna come down and whip Matt Carter if that damned order wasn't filled." That afternoon the Gypsy showed up again carrying the balance of the groceries. Matt had not known what to think of Maggie at first, but he learned that as long as she received what she requested there would be no problems—provided the goods were fresh. Once, he sent

Pictured in front of the logging camp cookhouse are members of the kitchen crew. Standing, top from left: Bernice Barnes, Maggie Biord-McNeill, Suzi Glass. Bottom from left: Ruby Barnes, Suzi Glass's daughter, James Biord. The logging camp was universally known as "Camp Maggie." (COURTESY OF SAM SWANLUND)

some rancid butter up to her, and she immediately phoned the store and screamed, "Matt, you'd better get out on those tracks 'cause that butter you sent up here was so damn rotten it just started walkin' back down the hill."

These outbursts were typical for Maggie. Her temper was short, and she made no pretense about it. But she was not entirely brimstone and fire; she had a personable side about her too. She cared for the people that worked for her, and if they passed her test of hard labor and dedication, they soon saw beyond her gruffness, and experienced a woman with a big heart.

Maggie separated from her first husband and continued to cook at the camp while raising her five-year-old son, James. She was joined in the kitchen by her sister, Suzi Glass, and the two of them worked together and helped to create a social life in the logging camp by hosting occasional card parties. Those were enjoyable get-togethers and the two sisters always concocted special dishes to enhance the evening's entertainment. But darkness entered Maggie's life when her son James was killed by an

The main logging camp, located 5 miles up the South Fork of the Elk River. The company established 5 different camps over the years in its forests above the town. This camp was the biggest and most permanent. (COURTESY OF EVELYN OBERDORF)

accidental gunshot wound just before his sixth birthday. Though burdened by this sadness, she was not one to complain of her misfortunes. Several years later she remarried and became Maggie McNeill, but up at the cookhouse she was always know as just plain Maggie.

One of the few times Maggie was ever embarrassed by her own antics was after an incident which occurred during her last reigning years at Falk. A friend of Noah's who had invested money in the company came to Falk to inspect the operations. During the course of his tour he happened to pass through Maggie's kitchen and the aromas caused him to lift a lid from one of the cooking pots. Seeing this nosey stranger at her stove, Maggie picked up a meat cleaver of considerable size and began running and swinging toward the man, yelling, "I don't allow no son-of-a-bitch to take the lids off my pots." After discovering the identity of the "son-of-a-bitch," Maggie felt a twinge of nervous embarrassment in the presence of the company's new investor. The incident was forgivable, however, because as everyone knew, that was Maggie.

One year, Maggie decided to take a leave of absence from her kitchen duties at the camp cookhouse, and the company was forced to hire a new cook. Whoever was to replace Maggie had a tough act to follow. A fellow by the name of Bowman was the first to try. His cooking was sloppy, though, and the food was greasy and tasted terrible. Serving bad food in a logging camp was as obvious as pumping water at the local gas station. After two weeks of this, the men began to grumble among themselves, but decided to give the new cook more time before confronting him with the issue. A week passed without any complaints, but the food did not improve. Finally something had to be done.

Peter Oberdorf was working in the blacksmith's shop and had stepped outside for a breather when he saw the woods crew walking down the tracks towards camp. Woods Boss Jim Copeland stood on the blacksmith's landing nearby and was nervously rolling the brim of his hat, which was a sign of trouble. "The goddam men want to run the ranch. Well, they can go ahead, I'm through," he said to Peter. "They're on strike for a new cook!" Copeland was upset because he had known nothing about the strike talk and, in the company's opinion as well as his own, he was supposed to be in close communication with the men.

The crew entered the cookhouse and two of the hook tenders negotiated for the strikers. It wasn't long before Winfield Wrigley arrived from town aboard the Gypsy. His mood was solemn but his message was clear. He flatly stated that those who did not return to work immediately were to come down to the office, collect their pay, and seek a job elsewhere. Winfield did not enjoy this position of enforcement, but as the acting company manager he had to take control. His position was somewhat strengthened when Jim Copeland stepped forward and gave his word that he would hire a new cook within a week. Satisfied that their demand had been met, the crew headed back to the woods.

The cook and his assistants had felt the pressure mounting and were ill at ease during the entire confrontation. When Jim Copeland told them that he was going to hire a new cook they became very angry. That afternoon the kitchen crew threw all the cookhouse food out the back door and down the hillside—meat, bread, milk, flour, everything, and departed on the next log train to town.

The company managed to quickly locate a replacement cook but his first day was difficult. He and his helpers had to scour the brush on the hillside for the discarded food in order to salvage the evening's meal. Things worked out better with the new kitchen recruits. The woods crew was satisfied and realized that not everyone could serve a meal like Maggie could. But with Bowman as a greasy reminder, they knew it could be a lot worse. The food strike was the only strike ever to occur at Falk.

Tales Of The Children

GROWING UP AROUND FALK WAS IDEAL FOR YOUNG-sters. They were able to take walks in the forest, find a favorite swimming hole on a hot summer's day, or tumble down the mill's sawdust pile that stood along the banks of the Elk River. Friends were always nearby to join in an adventure or just simple mischief. The latter usually culminated on Halloween night. On that particular evening, valley residents never knew what to expect from the pranksters who roamed the country roads. But the following morning, someone was likely to find their property re-arranged. The unlucky party might even have to search for their outhouse, for several times these devices were found teetering on nearby railroad trestles!

When the kids were not occupied with school, chores, or play, they had an opportunity to earn spending money during the summer months when the blackberries were in season. Whole days were spent filling one-gallon tins with the abundant wild berries. These containers were sold to the cookhouses for 35 cents each. Most of this cash was recirculated into the local economy via the general store's candy counter.

For the most part, the kids who lived in the town of Falk had a different life than those who were raised on the farms. Farm families relied on their youngsters to help make ends meet; consequently, these children had the most work expected of them. Arthur Forbes, for instance, a tree faller in the woods crew, also operated a farm in the Upper Valley with the help of his sons. Every day after school, young Ruben Forbes tended his father's cattle as they grazed the northern slopes of the valley. He spent entire afternoons combing the hills in search of strays, sometimes finding them over the divide in the Ryan Slough area. When the cattle were accounted for, Ruben would then begin the barnyard chores. In addition to these responsibilities, some of the farm children had to walk four miles round trip every day to attend the Jones Prairie School.

Thinking back over the years, several former residents of the Upper Valley recalled some of the memorable events of their childhood. From these remembrances come "Tales Of The Children."

When Leland Newell was a young boy, his mother planned a trip to San Francisco by ship. The thought of this trip caused Leland much anxiety. He had never been on a boat before, much less out in the ocean, but his mother reassured him that everything would be fine. He stuck close by her as they boarded the *City of Topeka*, a lumber-passenger schooner that frequented Humboldt Bay. When the lines were cast off and the schooner began chugging down the bay, Leland thought that being on board a ship was not as bad as he had imagined. He stood near the rail, watching the wake from the schooner's hull, and began to feel quite relaxed. He was suddenly caught off guard, however, when the *Topeka* turned hard west and plowed into several large waves as it crossed the Humboldt Bay harbor entrance.

They had been told it would be a stormy voyage and the captain had requested that all passengers stay off the decks. Leland spent most of the trip in his stateroom, getting only partial value for the $14 round trip, since he couldn't hold down the meals that were included in the price. The passage around Cape Mendocino was the most violent part of the journey. Large swells crested and broke over the decks of the *Topeka*. Scared to death and suffering from seasickness, Leland thought the two day trip to San Francisco was a lot longer. Although they arrived with no mishaps, he didn't enjoy his time in the city, because in the back of his mind lurked the fearful images of the return voyage.

Nevertheless, on departure day, he once again boarded the *Topeka* and hoped for the best. Luck was not with poor Leland on this trip either. The return passage proved to be as nasty as the first. To make matters worse, when they rounded the treacherous waters of Cape Mendocino, the *Topeka* struck a huge log that was awash in the storm. Leland was convinced that the ship was sinking. The *City of Topeka* did make it to port for repairs, which proved to be minor, but young Leland made a promise never to journey by ship again. It was a promise he kept. The next time his family went to San Francisco he stayed at home rather than risk another adventure aboard a leaky old schooner.

⌐ɯɯɯɯ ̣

Ruth Christie's family moved into Eureka to live for a couple of years but her father, Herbert, remained the boss of the choppers in Falk's woods. It was too great a distance to commute every night after work, so he kept a cabin near the mill and came home on the weekends. During the summer, when Ruth was not in school, she visited Falk and spent a week at a time with her Uncle Ulysses and Aunt Code, and her father came by to see them in the evenings after work. Ruth always anticipated his arrival. She sat on his knee and talked about whatever needed talking about, but the greatest enjoyment was just to be with her father.

When Ruth stayed at her aunt and uncle's home she slept in the guest room which was across from the parlor. Her room was cheerful and the bed was covered with a handmade down comforter that was soft and warm. But the parlor haunted her. Ulysses had been a hunter and trapper for years, and it was in this room that his stuffed trophies were displayed. The room had a very mysterious feel about it. The curtains were always drawn over the windows, and no one was allowed to sit in there unless it was a special occasion, such as a visit from the Parson. From the hallway, Ruth could peer into the dark parlor and see the motionless figures that stood in the shadows. It was quite a frightening sight to her, but even when she was in the house alone, her curiosity compelled her to sneak glimpses of the room and its silent creatures. Some were crouched with open jaws; others had sharp beaks and spread wings. To Ruth these frozen images could have come alive any second.

Bedtime was especially frightening because she had to pass through the parlor to reach the safety of her room. There was an enormous great horned owl that stood near her bedroom door, and it was almost as tall as Ruth. She would bolt past the bird, but she could never avoid its angry stare, so she always made sure the door was shut tightly before she hastened into bed. Once under the down comforter, though, her fearful visions vanished, even the angry eyes of the owl.

⌐ɯɯɯɯ ̣

The James McManus family was away from their farm one Sunday morning and their nine-year-old son, Albert, was left behind to "hold down the ranch." Albert took his job very seriously, and was alarmed when the watch dogs began barking fiercely outside. He grabbed his .22 rifle and went to investigate. He had no sooner stepped outside when he was confronted by a black bear and her cub. Not being old enough to recognize how dangerous the situation was, Albert started shooting. The cub was killed, and the mother bear took flight and escaped to the woods. He had quite a story to tell when his folks got home that afternoon.

During a recess squabble, one of the young boys at Jones Prairie School pushed a girl who had been teasing him, and she ran to tell the teacher, Maude Frost. Maude had a reputation for not tolerating any nonsense, and when she called for Harold Weatherby to remain after class everyone knew he was in for some trouble. And trouble it was. Maude whipped him with a leather strap. The punishment was somewhat severe for the incident, and Harold was not going to let her get away with this. He planned revenge.

Tearing away a board on the back of the women's outhouse, he lay in the bushes with a large bunch of stinging nettles, waiting for his victim. When Maude came to the privy and took her seat, he gave her a swat with the nettles that she, also, never forgot. As Harold ran away, sweet revenge began healing his welts, and the hurt pride that had swelled with them.

Outside Charles Steenfott's home there existed a huge sawdust fill that the children played in. One day his son, Norton, and an older cousin were out on the pile with Norton's wagon. Norton used to ride the beat-up old wagon down the steep slope of sawdust, and, upon reaching the bottom, always had to hammer the wheels on tight since they were loosely held on with nails. For this task he carried his mother's potato masher in his belt.

Norton's cousin wanted to try a run on the sled, so down he rode, but when he reached the bottom, he began walking back up without the wagon, despite Norton's yelling for him to drag it up. Norton had to climb down to retrieve it, drag it up the hill and pound the wheels on, only to hear his cousin say he wanted to go again. Norton reluctantly obliged him, but the same thing happened again. Norton had to go down a second time and retrieve the wagon, pound the wheels on, and drag it back to the top of the sawdust pile.

The cousin wanted to go a third time but Norton said "no way." When the selfish lad piled into the wagon anyway, he was met by a blow to the nose with Norton's potato masher. There was blood everywhere. The cousin ran off screaming and poor Norton didn't know what to do. So he hid under the railroad trestle. All afternoon his mother called for him. At one point, she was standing right over him on the trestle while she called. But Norton was wary and was not going to come out for anything. His mother called out that she was going to go pick blackberries, but he still would not move. As she walked away she called one last time for him to look out for the wildcat.

Norton's mother and father used to tell him that a wildcat lived down by the trestle so that he would not wander past that point. Not wishing to encounter the wildcat, Norton scooted out from under the trestle and ran along behind the bushes, following his mother as she headed to the berry patch. This went on for a distance until she finally said, "Now you come out of those bushes and quit your hiding. Take this bucket and start picking!" Norton scrambled out of the bushes and picked for all he was worth!

Chester Barnes and a few of his friends had decided that the steep slope above the schoolhouse could be made into a perfect sled run. They were all enthusiastic about the adventure. The only item they lacked was the sled. To solve this, the industrious Chester procured the needed wood materials from the mill workers, who, remembering Chester's Saturday night dance concession, figured this plan would also be worth supporting.

He and his friends spent several days constructing a very large sled which was big enough for everyone to ride on at once. The runners were made of hardwood and were greased to reduce the sled's friction. The crew of kids then cut a path through the bushes for their planned descent, and finally they were ready.

A dozen youngsters dragged the 10-foot-long sled up the hill to the summit and everyone climbed aboard for the fun ride down. No one had any idea what lay in store for them. They pushed off and began to descend much faster than any of them had planned. Down they came, almost immediately leaving the path they had cut, tearing through bushes, careening wildly and accelerating the whole time. Kids started to bounce off, and some jumped clear. When the sled came to rest upside down at the bottom of the hill, the stunned riders were strewn

A group of children take time out from play to pose for a picture. Top from left: Herbert Newell, Neil Crowley, Leland Newell holding a black and white dog. Bottom from left: Lawrence Crowley, Katherine Nellist, Merced Wrigley, Shirley Callihan, Evadne Miller. (COURTESY OF EVADNE BRADLEY)

haphazardly along the path, two with broken arms and others with scrapes and bruises. This event was talked about for quite some time. When Chester was asked about the ride, he just said, "she was flying."

⌐ᴐᴐᴐᴐᴐᴐ⌐

Evan Rushing, Jr. rose early one Saturday morning to walk to Eureka with the intention of buying some new shoes. With $2.50 in his pocket, 10-year-old Evan felt pretty lucky that day as he headed into the city on the railroad tracks. His family was poor, and the one pair of overalls he owned had quite a few patches, so this was a major purchase he was intending. After eight miles of walking, Evan finally reached Broadway in Eureka, where he ran into Jet Christie, who was sitting in front of the blacksmith's stables. Jet was also from Falk; in fact, he and Ev were related. Jet worked on the same woods crew as Ev's father and he had quite a reputation about town as a heavy drinker. According to all accounts, he managed to live up to it quite convincingly. Drinks were ten cents in those days and he often asked passersby for a dime. For that price you usually got a first-rate story, even if it was invented at that very moment.

Jet saw young Ev coming down the street and asked where he was going. Ev explained that he was in town to buy a pair of shoes. "How'd you like to buy a horse?," Jet asked. Ev knew how much a horse was worth, and figured he did not have enough money to buy one. But Jet took him over to look at a little bay that was in a stall. "How much money do you have?" Ev said he had $2.50. "That'll do it, just give me the money and it's your horse," Jet replied.

Evan was tickled to death. He gladly gave Jet the money and was going to get his horse, but Jet explained that the horse was waiting to be shod and would be ready when the blacksmith returned. Ev walked around town for a couple of hours, thrilled with the deal he had struck, although he did not have a penny left for shoes. Later in the afternoon, he stopped by the stable to pick up his horse. He walked into the blacksmith's shop and asked for the bay.

"What horse?," the blacksmith said. "You haven't got one here." Ev explained that he had paid cash for Jet Christie's bay. The blacksmith split his side laughing. He told Ev that Jet was probably out in Freshwater by then, having one "heckuva gay ole time." Ev walked the eight miles back to Falk with no horse, no new shoes, and no money. When he returned home, his father was furious over the incident. The next day at work, Jet was forced to quickly come up with $2.50.

DECLINE

Previous page, an empty cookhouse. (COURTESY OF PETER PALMQUIST)

*". . . It was a model community; its citizens,
drawn from the four corners of the earth, were the most friendly
and amiable fellows I had thus far encountered in my travels . . "*

CAPTAIN SMALE, PRUSSIAN SEAFARER AND AUTHOR.

Passing
The Torch

UNDER NOAH FALK'S DIRECTION, THE ELK RIVER MILL and Lumber Company became recognized as one of the mainstays in Humboldt County's timber industry. By 1920, however, rumors were circulating that Noah was considering selling the company. This news was taken in stride by most people. A change in ownership would probably have little effect on their lives so long as the business continued to flourish.

The rumors had begun after Charles Falk's wife, Blanche, had suddenly become ill. During a critical operation, the young woman had died. The family was stunned, as were all those who had known her. People came from all over the county to Noah's home, where the funeral took place. The days were very solemn around the Falk family during that time. Blanche's death lent credence to the talk of a possible sale. Neither Charles nor his father lived in the town of Falk, and everyone knew Noah was well on to 85 years old. The time had come to pass the torch.

In the summer of 1920, nearly 40 years of inspirational work culminated when Noah sold the company and all of its properties to J. R. Hanify of San Francisco. Noah and Charles were the last of the Falks to leave the Elk River Mill.

Hanify had been watching timber developments around the Elk River for quite some time. Beginning in 1901 he and a partner, Albert Hooper, bought half of the Bucksport and Elk River Railroad's stock. The partners also purchased over 6,000 acres of redwood timberland in the valley. When Hanify gained title to the Elk River Mill and Lumber

Noah Falk at age 90. The hardy lumberman was photographed during an interview granted several months before his death. (COURTESY OF OREGON HISTORICAL SOCIETY)

Company, he also purchased Hooper's interest in the railroad. To further enhance his new company's independence, he bought three steam schooners with which to haul the mill's lumber products. Hanify's investments seemed to give a new boost to the little town's future.

All seemed to be going well for the enterprising San Franciscan until he was involved in an ill-fated boating accident and drowned in San Francisco Bay. This mishap occurred only a short time after he had taken over the mill. All of his investments were left to family heirs, who subsequently set up the J. R. Hanify Company to manage his interests.

Soon after Hanify's death, a former associate of his, John Reede, bought the Elk River Company. Reede, also from San Francisco, continued to base the company's ownership in the big city, but allowed Winfield Wrigley to direct its operations. Winfield, who was also a minor partner in the firm, became acting president. Even though the company had gone through two complete changes of ownership, life in the town of Falk remained the same.

NOAH FALK, 1836-1928

On March 10, 1928, Noah Falk died at his home in Arcata at the age of 91. His passing was met with a great deal of community respect and admiration for a man who had witnessed and helped forge nearly a century of change. The contributions he had made to the betterment of people's lives was his life's work. Over the years he created hundreds of opportunities for people who had come to settle and work in the redwood forests of Humboldt County.

After Noah sold the mill at Elk River most of his remaining land holdings and business interests were in the Arcata area. In retirement, Noah and Nancy had been spending most of their time at their permanent residence in Arcata after living in Falk for nearly 30 years. They were a well-known couple in the county and led an active social life. In 1913 when the movie *Valley of the Giants* was being filmed in Humboldt County, Noah had been the film's technical advisor for the logging scenes. It was a role he was well suited to, since the film was about a man who made his fortune logging the redwoods. He spent several weeks on location with the film crews and later he and Nancy gave a party for the entire cast. During the party Noah was asked about the true location of the valley on which the film's story was based. He laughed and explained that the real valley was near Ryan Slough in Eureka, but that the film crews found the Mad River area more adaptable for their sets.

Noah kept his hand in business even after he sold the mill. He was a director of both the Arcata Bank and Eureka's Humboldt County

Bank, and owned a partial interest in the Arcata Hotel, where he liked to sit in the lobby during the afternoon and visit with his friends and neighbors. As a gesture of goodwill and thanks, one of Noah's last community projects was to personally donate a new fire engine to the city of Arcata.

He was a tough oldtimer, active even in his later years. He might have lived longer were it not for an unfortunate accident that occurred when he was 90 years old. While crossing an intersection in downtown Arcata one day, Noah was struck by an automobile. Although he achieved a partial recovery, the hearty lumberman never fully regained his physical health. His mental capacities were quite sharp however, and allowed him to continue with his community interests, and also complete a lengthy interview for a lumber journal from Portland, Oregon, in 1927.

Noah and Nancy had been married 66 years when he passed away. Of the original Falk family, only Elijah and his sister Elizabeth remained, surviving Jonas, who had died in 1917. But the Falk name would be carried on for generations, by his son Charles, who had worked with Noah for 30 years, and by a multitude of nieces, nephews and grandchildren. With Noah's passing, one of the true giants of the Elk River valley had been laid to rest.

CHANGES

During the 1920s the company's new ownership energetically pursued business as usual. The mill whistles still blew, the trains rolled, the school bell rang, and the dangerous work of redwood logging continued unabated. But events were changing working conditions in the woods. The state had recently passed laws to protect people in dangerous occupations. Yet when Evan Rushing was injured in a logging accident, he found himself embroiled in a legal action that developed between the state, which was trying to enforce the new laws, and the company, which was reluctant in complying with them.

Evan had grown up in Falk, and had worked in the woods for over ten years. He was recognized by those around him as one of the most capable men in the crew. Evan had been fortunate during his years of logging, and though he had seen men carried out of the woods with injuries, some never to return, he had never been the victim. Part of this good fortune was due to skill and alertness, but to survive in the woods a certain amount of luck was also needed. On one particular day Evan's luck ran thin.

The logging crews had recently entered a new creek basin and were involved in removing the area's timber. Evan, who was the zoogler, had finished securing the head rigging onto another log set while his brother

Evan Rushing, Ivan Christie and Lloyd Rushing (l. to r.) pose for a picture before they ride a line of logs down to the landing. (COURTESY OF EVAN RUSHING)

Lloyd signaled the bull donkey engineer to start the winches. The bull's cable pulled tight and the Rushings were on their way down the main chute with another string of redwood logs in tow. With over 20 years of experience between them, this was a routine operation for the brothers. While the logs moved down the chute, Lloyd slung water and Ev kept watch on the rigging and couplings, as well as the nearby haulback line. There was no reason to think that this particular run would be any different than the hundreds of others they had overseen.

About a quarter mile from the log landing, Ev suddenly noticed that the haulback line had pulled out of its spool, and the cable had become snagged on a bridge. It was not unusual for the cable to snag like this. Ev grabbed his double-bitted axe, jumped off the moving log set, and rushed toward the bridge. Several swings of the axe was all it took to free the cable. When it came loose, however, it whipped out of position and solidly hooked Ev's foot. The fast moving, high-tension wire rope snatched him from the bridge and flung him, end over end, 30 feet in the air. While in mid-flight Ev managed to throw his axe clear, but he landed beneath the bridge with one leg jammed against an old stump.

Lloyd saw the accident, and after signaling the bull donkey engineer to stop the cable, ran to help his brother. He climbed down the steep hillside and found Ev lying in the creek below. He was still conscious but his ankle had been severely injured and he was in tremendous pain. Lloyd carried him out of the ravine and placed him on the lead log, where he lay until the bull donkey had winched them down to the landing. The injured man was put aboard the log train and rode down the canyon to town, where he was helped to the company office.

Despite his request, the company did not feel obligated to drive Evan to town, since he was still breathing and all of his body was intact. He ended up hitching a ride with a traveling salesman who insisted on rushing him to the hospital emergency room after seeing the injury. The ankle was so damaged that Ev could not walk for three months, and a year later he still limped.

The hospital filed a report with the new State Compensation Board for industrial accidents. This board had been created after passage of the Workmen's Compensation Insurance and Safety Act in 1913. Companies were now held responsible for most on-the-job accidents, and were expected to contribute to a state-wide worker relief fund. The state board requested that Evan appear for a hearing and a referee was sent to Eureka to review the case. After hearing both sides of the story, it was determined that the Elk River Mill and Lumber Company was liable for the accident, and was ordered to pay a $1,000 settlement.

When Ev stopped by the mill office to see about this matter, Winfield Wrigley told Evan he didn't have enough of a case to claim the money. Ev did not know what to think. He had not filed the claim himself, and obviously the injury was not intentional. He felt caught between the new state agency and his own company.

One month later the state contacted Ev to see if he had received his payment. When he said no, the state began to pressure the company. An ultimatum was issued. For every day the debt was not paid, a 25 percent interest rate would compound. Three days later Ev received his compensation. Retribution, however, was not long in coming.

During the following spring, when Falk's woods began rehiring, Jim Copeland asked Ev to come back and work for him. "They're not going to hire me after my injury claim," Ev told Jim. "I run the hiring," replied Copeland, "and I know the story of your accident, so you're hired."

Evan went back to work in the woods, and later in the summer he was asked to engineer the log train while Jesse Barnes took a two-week vacation. Since he was familiar with the No. 2 Falk engine, Ev agreed. The following week the new owner of the company, John Reede, happened to come up from San Francisco to inspect his holdings. One afternoon he climbed aboard the locomotive that Ev was operating, took a careful look at him, said a few brief words, and coolly departed. Reede, having recognized the injured claimant, went straight to Winfield and ordered Evan fired.

When Jesse Barnes returned from his vacation a troubled Jim Copeland stopped Ev as he started to board the morning crew train. "I hate to tell you this," the woods boss said, "but John Reede says you have to leave. I want you to come up and have breakfast with us before you do." Later that morning Copeland shared his disappointment with Ted Barnes. "I feel terrible," Jim told Ted, "I had to let one of the best woodsmen I've ever had go this morning."

Evan rode the crew train to Camp Maggie and had one last meal with his fellow workers, some of whom he had known his entire life. After breakfast he bid his farewells and rode the train back to town. Later that morning Evan Rushing packed his belongings and left the Elk River Valley.

Last
Hurrahs

THE 1930S USHERED IN FAR-REACHING CHANGES. DURING the previous decade, America had been free from war and the economy had flourished. But these thriving times reached the summit of prosperity, and people were forced to hold on tight when life's roller coaster began its descent. Many of the businesses that had been created in past years became distressed. Unemployment swept the nation. This was the beginning of The Great Depression, and the town of Falk was shaken by the force of this economic disaster.

By the summer of 1930, several Humboldt County mills had already shut down production, but the Elk River plant limped along. In early October, however, word was received from John Reede that the mill and logging operations at Falk were to be halted until further notice. Within hours of that phone call all machines at the Elk River plant ceased to operate. The town lay quiet. Single crewmen were the first to leave, but most of the families remained in the hope that the company's operations would resume. Winter approached, the daylight hours grew short, the rains began to pour, and at that point the town seemed lifeless. Once again, nature and circumstances were testing the will of the small forest community.

Although the mill closing created a solemn mood in the valley, some optimism lingered. The Bucksport and Elk River Railroad Company had decided to make vast improvements on their line, and in 1931 rebuilt the entire track. The company also extended the line up the North Fork of Elk River, allowing the Dolbeer and Carson Lumber Company to reach some long isolated tracts of timber. During the following year a complex series of business transactions formed a new company, known as the Bucksport and Elk River Railway. Essentially, the line had been rebuilt, refinanced, and reorganized, giving the Dolbeer and Carson Lumber Company complete ownership. Two years later the Trinidad and Bucksport engines were sold for scrap, and Dolbeer and Carson's larger, more powerful locomotives were brought in to haul the loads of logs and lumber from the Upper Valley.

Despite the shut down the Elk River Mill continued to ship small amounts of lumber from the huge stockpiles that had accumulated on its storage decks. A skeleton crew of a dozen men, headed by Winfield Wrigley, was employed to move this lumber, and also to maintain and watch over the company's properties.

The town idled along in this manner for several years, during which time its people lived frugally and continued their hope for a revitalized livelihood. Former residents often came back to visit their friends and families. But, with the mill closed, people had few chances to work and the town edged slowly toward abandonment.

In the summer of 1933 Norton Steenfott and a partner drove to Falk to visit an old acquaintance of theirs who still had a job at the mill. Norton had grown up in the town, but had not lived there for many years. It was a slow trip into the valley, and the two men consumed a fair amount of wine during their drive. When they arrived in town they were full of the "Dutch courage," and set out to locate their friend. Norton thought the man lived across the river near the dance hall, so he told his companion that they would drive there. "What da' you mean, drive there," his friend replied, "there's no road." Norton simply ran his car up onto the railroad tracks, straddled the iron rails, and set off bumping along toward the trestle that crossed the river gorge. They stopped the car in the middle of the trestle and looked out the windows at the river below, enjoyed a good laugh, and then without hesitation continued across to the other side.

A few workers at the mill had stopped to watch the vehicle negotiate the narrow train trestle. Norton, who was sure his friend lived near the dance hall, stepped from the car. He yelled across the river to the men at the mill, asking where his friend lived. Amidst some chuckles, they pointed to a small cabin above the mill on their side of the river. "No problem," said Norton, "I'll just drive back over the trestle to the mill."

It was at this point he realized his predicament. There was nowhere he could turn his car around. The tracks on his side of the river were laid on a steep, six-foot high railroad bed and a parked rail car blocked the track beyond where he had stopped. Norton's sense of euphoria quickly vanished when he realized he would have to back his vehicle across the trestle. Upon

Falk's two locomotives sit idle after being abandoned on a side rail near town. The No. 3 Heisler is in front; the No. 2 Baldwin is behind it. (COURTESY OF HENRY SORENSEN)

hearing the plan, Norton's partner quickly abandoned him, and set out on foot for the other side. Norton hung out the open door and watched the wheels as he backed and bounced the car across the rail ties, one at a time. This took a great deal of concentration, especially when crossing the river gorge. Norton, who was now cold sober, managed to maneuver his vehicle safely off the tracks. When he finally found the man he had been searching for, he was ready to drink some more wine!

Even though the company was shut down indefinitely and there was little activity in the town, Winfield and Grace Wrigley remained in Falk. They lived a quiet life in Noah's old house overlooking the mill. Both of their children, Merced and Dorothy, were away at college, but at Christmas time in 1934 Merced came home for a visit and told his parents that a young lady he had met in San Francisco would soon be arriving by train to meet them.

Karla Brewer had been raised near San Francisco and was used to city living. In spite of a torrential downpour during her train ride north, Karla was captured by the beauty and remoteness of the redwood forests. She also thought Eureka was a quaint little community as Merced drove her from the train station. But when they left the city limits and wheeled toward Elk River, she began to wonder where the narrow gravel road was leading them.

For quite some time they drove through the rainy night. The road surface changed from gravel to wooden planks, and while they thumped along in the dark, the small outline of dim, rustic cabins began to appear, marking the outskirts of Falk. The town was dark, without electricity since the mill's generator shut down. Karla began to suspect that Merced's family were hillbillies. But when they pulled up in front of the family home, Grace came down from the front porch with a coal oil lamp and gave Karla such a warm greeting that all of her unsure thoughts vanished. Three months later Merced and Karla were married.

HARD TIMES

For a person who was not familiar with rural life, nor the hard times in the timber industry, the little wooden shacks and the children in patched clothing were visible signs of poverty. But while things were lean in Falk, the people there were much better off then the unemployed in the cities. The townspeople could at least put food on the table by tending their gardens, fishing and hunting, all the while nurturing their hopes that the mill would reopen.

But the Depression was relentless in its effect on working people. Because of the unemployment situation, frustrations were running high throughout the county. Some mills had reopened with limited crews, but there had been confrontations among the workers themselves. There were strikes and strike-breakers, threats and vigilante action, and charges of communist agitation. On June 21, 1935, these tensions came to a head at the gates of the Holmes-Eureka Mill, when 200 timbermen fought with the police. After the tear gas and crowds had cleared, the casualties were added up. Three men had died as a result of the violence and scores were injured, including five policemen. The carnage might have been even worse. At one point, a Gatling gun was turned on the crowd, but luckily it jammed, preventing what could have been a massacre.

The riot placed a dark shadow over the county's timber industry, but the following year a ray of light broke through for the people of Falk. In May 1936, the Elk River Mill and Lumber Company reopened the entire operation. The company announced that it would hire only those who had been previously employed in the town. This brought welcome relief for the people who had remained in the Upper Valley during the last five years.

Much of the equipment in the woods and mill was now outdated, and a substantial amount of cash was needed to replace the machines. Winfield Wrigley was a shrewd businessman, and was not afraid to borrow large sums of money when he had confidence in his plans. Through the company, he secured a $1 million loan from the federal government, and refinanced the entire operation. When the first crews began to cut timber for the mill to process, they discovered big changes had occurred in the woods. The steam donkeys had been replaced by two metal-tracked crawlers which could claw their way up and down steep hillsides and move the logs to the landing.

These tractors proved their efficiency, but the old-timers grumbled about something being lost, and many of the crew jobs were eliminated with the machine's success.

But the old-timers were pleased when one old veteran was lifted into a place of honor. In 1936 the Elk River Mill and Lumber Company removed the little Gypsy locomotive from its graveyard siding and gave it to the City of Eureka as a gift for the annual Fourth of July parade. After a complete restoration, the Gypsy entered the parade with flags flying, and chugged its way up the streetcar tracks in Eureka. The little engine had earned its right, beyond any doubt, to be placed on permanent display at the Fort Humboldt Historical Park. It was the oldest surviving locomotive in the area.

One year later, all of the town's hope for prosperity was snatched away. During the summer of 1937, word was received that all operations at Falk were to shut down again, this time, for good. The resurgence in the timber industry had proven to be short-lived. Ted Barnes was working in the woods the day the news was broken. At noon the woods boss, Bill Fleckenstein, walked up to him and said, "It's all over, Ted, they're closing it down." With those words the crew left all the equipment in the woods, rode the last log train down to the town, and picked up their pay checks at the office. The two locomotives were parked on a siding and the steam engine at the mill was shut down. This equipment never operated again.

A few long-time employees remained on the Elk River Mill's payroll until the last of the lumber products were shipped. Red Braghetta, who had spent nearly 20 years in the front office, worked closely with Winfield during this process, and he also took care of finalizing the company records.

The closure had become inevitable due to high operating costs and low shipping volumes. The mill had managed to move only 649 cars of lumber during the entire year. In the past, that would have been just one month's tally. When these facts surfaced in the ledgers, John Reede contacted Winfield and gave the final command. One telephone call was all it took. Within weeks of the closure most of the residents had moved away. Falk began to take on the look of a ghost town.

For those who remained in the outlying areas, other choices were available. The automobile was now a common form of transportation and gasoline was selling for 16 cents a gallon. People were able to live on their rural homesteads and commute to more distant jobs. North of Falk a development was occurring that offered some opportunity. The Dolbeer and Carson Lumber Company had established a logging camp on the North Fork of Elk River called Camp Carson. They had been hauling timber out of the canyon for three years and some of the men from Falk found work there. But because no sawmill existed only woodsmen were needed.

Winfield Wrigley stands on the empty lumber decks of the Elk River Mill during his final days in the town of Falk. Already the brush and trees are beginning to obscure the townsite as Winfield's house, which had once been Noah's, is partially hidden from view. (COURTESY OF DOROTHY WRIGLEY)

One of the men who worked on the North Fork project was Ted Barnes. Ted had gained a good deal of logging experience at Falk and when the mill shut down he simply went over the hill to Camp Carson. In the early spring, during a log removal operation, Ted got tangled in an out-of-control wire rope and was launched high into the air. When the confusion was over, he took a forced vacation for several months, which he spent in a partial body cast while his broken bones mended. Towards the end of the logging season, following his recovery from the accident, Ted decided to attempt a little work before the winter rains came. On his first day back in the woods he and his partner were running from a wind-toppled fir snag, when his partner blindly threw his axe. The ill-thrown axe lodged itself in Ted's knee. This accident was compounded when the wound became infected with redwood poisoning. Ted felt quite discouraged when he had to spend the next few months in bed with another cast. At this point he considered changing occupations.

Luckily, all of Ted's experiences at Camp Carson were not detrimental to his health. During his time at the North Fork operation, Ted met and married Mary McKee, who worked in the logging camp cookhouse. They moved from the Elk River Valley and settled in Eureka. Ted gave up logging and began driving freight trucks. He had gotten out of the woods with his life, and that was better fortune than many men had known.

At the same time that Ted had decided to leave Carson's woods, an elderly acquaintance of his from Falk arrived at the camp. Jack Miller had been the blacksmith at Falk for 32 years. When the company closed, someone was needed to look after the valuable machinery and tools that remained in the buildings, and Jack had become the watchman. He was recognized in the county as a top-rate blacksmith, and when the Elk River Mill shut down he had received many offers of work. He declined these offers to remain in Falk, but the watchman's job seemed to him a retirement position, and he missed the challenge of blacksmithing.

Jack's son, Wayne, was employed at Camp Carson as a woodsman. Carson's blacksmith had decided to leave, and the crew boss asked Wayne if his father might be interested in the job. Jack was interested, but only on one condition: he did not want to take the job away from another man. When that point was clear, Jack began blacksmithing at Camp Carson. He was over 65 years of age.

Three years later, when Wayne was working in the woods, one of the camp bosses came to tell him the grim news. His father had been working by himself on a strenuous project in the company shop, and had suffered a fatal heart attack. Jack Miller had lived and died by his work.

By 1940, only a dozen people remained in Falk. Most of the structures were now abandoned. People had traveled lightly during the exodus, leaving behind stoves, bathtubs, beds, furniture, dishes and even clothing, as if to record how quickly one's idea of permanence can change. Brush and young trees sprouted up around the vacant buildings. Slowly, inevitably, the patient forest began to reclaim the town.

Aftermath

NATURE WAS RE-ESTABLISHING ITS DOMINANCE AT Falk, but it was still held at bay by a few hardy survivors who continued to live in the town. Most of these folks were older men who had no families elsewhere, no retirement pension, and were too old to compete in the job market. These men lived close to the land, looked out for one another and existed in whatever manner they could. The company's presence had not totally disappeared either, and in 1941 the final stand was made.

Winfield, who had been the driving force since Noah had stepped aside, was intent on repaying the $1 million federal loan. With that goal in mind he made a move that eventually saved the Elk River Mill Company from bankruptcy. A substantial redwood forest still stood on the far reaches of company land, and Winfield contracted to sell this timber to the Hammond Mill. A private logging company owned by Bill Hess won the bid to fall and deliver the logs. Hess established the first logging road in the valley and removed the trees from the distant hills by truck. Three years later, Hammond's logging crews cut the last of the timber and hauled it to the mill in Eureka. The Elk River Lumber Company realized a profit of only $4.00 for every 1,000 board feet of harvested logs. But this small profit, along with the eventual sale of the company's land holdings, enabled Winfield to completely pay back the million dollar federal loan.

After Winfield had hired Hess, he and Grace left the big house by the mill and moved to Eureka. The lumber had been sold and the logging operation was independent, so his presence was no longer needed.

Winfield handed the keys of the town to Red Braghetta. Red and his wife Ruth, who was Winfield's youngest sister, inherited the caretaker's duties. Red was one of only two people who remained on the Elk River Lumber Company's payroll. The other person was Frank Falor, the wharf boss, who had stayed with the company during the six year closure, without wages. For doing that, he was allowed to keep his house for life, and through a company-established trust fund, received a retirement wage over a 25 year period.

The general store became Red's office and, although it was no longer stocked with goods, Red always kept a fire burning in the wood stove and welcomed friends to stop by to visit, play cards, or just have a beer. The store also had the only telephone around, so in its fine tradition the general store remained the town's social center.

One of Red's jobs as caretaker was to collect the rents from the few people who remained. In 1941 a two-bedroom home in Falk cost $5 a month. This fee was more a formality than an incoming cash flow, and if one of the oldtimers was hard pressed for money the rent was overlooked.

Doc Van Warner never paid rent. When the mill shut down Doc had packed his belongings, walked up onto a hill above town, and built a homestead. He was a self-educated, articulate, soft-spoken man who had decided to spend the rest of his days in seclusion. He tended a large garden and raised rabbits which supplied most of his food, and also fished and picked berries. Doc, at 70 years of age, rode his bicycle into Eureka once a week with a box full of rabbits strapped to his back. The sale of the rabbits financed his purchase of staple foods. He also canned fruits and vegetables, made pickles, sauerkraut, venison jerky, and home brew. His garden area was about 60 feet in diameter and was fenced in with large redwood slabs which he had split and sunk into the ground. This sturdy wall protected his vegetables from the deer and other wild creatures that roamed the hills above Elk River. Doc had seen many sides of life, and he readily shared his views and opinions in conversations with people who came to visit him. But to reach his homesite, one had to walk a two-mile foot path up the McCloud Creek Basin. Since he was so isolated, few ever found his home, but for those who did, a rich experience awaited them. Doc remained on his homestead until 1946, when he moved to a rest home in Eureka, where he died two years later.

Life in Falk became increasingly sparse, and even the oldtimers began to move on. For this reason it came as a great surprise that anyone would want to move into a town that had no future. But in 1941 a young couple rented a house down by the river behind the general store. Ed and Dorothy Erickson had little money, and in their struggle to make a living

Bandsaw blades strewn about the abandoned filers' shop. When the mill was operating, a team of men was employed just to keep the blades sharpened. (Courtesy of Dennis Sullivan)

Ed had found employment with Bill Hess's logging operation above town. Their story echoed that of the hundreds before them who, since 1882, had traveled up the Elk River Valley to find a home and work in the woods. The difference was that the earlier settlers had helped nurture the growth of a community, whereas Ed and Dorothy were witnessing only the hollow shell that remained of that growth. Within that shell, however, the spirit of the community still lived.

Ed and Dorothy were soon joined by Ed's brother-in-law, Ernie Fleckenstein, who also began working at Hess's logging site. The three young newcomers quickly adapted to their environment, and as they came to know the older men in town they helped them whenever possible. Ed and Ernie began regular hunting trips into the woods, and provided much of the fresh meat for the elders. A few years before, a bull had escaped from a valley rancher and roamed far up the canyon. It occasionally bellowed in the evenings and, after some time, managed to lure several cows from the fields into the forests. A small herd of wild cattle was the result. Ed and Ernie often went up the river to hunt one of these beasts, and carried the meat back to town. The old men were always very thankful, because it meant there would be plenty to eat. Red Braghetta discovered that the wild old bull was sometimes a dangerous critter to confront. While walking in the hills one day, he met face to face with this huge animal and after a momentary standoff, Red clambered up an old stump and spent the rest of the day sitting out of reach of the angry bull's horns.

As the town's human life slowed to a trickle, the wildlife in the woods around Falk became more abundant. Raccoons were seen by the dozens, civit cats raided the chicken coops, deer grazed among the empty buildings, and several bears moved closer to the town.

One day while walking along the railroad tracks above the mill, Dorothy and Ed witnessed one of these creatures at play. A black bear was up an old limbless snag, and having no idea that he was being watched, the bear slid down the tree, digging in its claws, making quite a racket. Once at the bottom of the snag the bear quickly shinnied back to the top, and amid a clamor of ecstatic grunts and scratching sounds, proceeded to come down again. The bear repeated his performance several more times before he finally tired of the game, and moved on to more important business.

Dorothy, for the most part, enjoyed her time in Falk because of its quiet beauty. Also, her neighbors added a good feeling to her life, but since no other young friends lived there, and Ed and Ernie were away working most of the time, she felt too isolated from people. One particular evening Dorothy had a trying experience.

It was on a Sunday, and Ed and Ernie had decided to walk up to visit Doc Van Warner to see how he was faring. Dorothy, who was pregnant at the time, decided to remain at home. The two men promised they would return by nightfall. After they left, the light breeze that had been blowing increased to a strong wind that rocked the trees, violently pitching them back and forth. Limbs began to break and fall everywhere. When darkness came the two men still had not returned. While she was home alone Dorothy became very nervous, as several limbs crashed onto the roof of her house. Falk was a lonely place at a time like this. Her anxiety soon got the best of her, and she went to tell Red of her fears. Red located several old-timers at their cabins, and the group set off with lanterns to search up McCloud Creek for the missing men. While the winds continued to whip the forest, the search party moved through the night as cautiously as it could until the men reached Doc's cabin. From inside the cabin the men heard what sounded like loud laughter. Upon reaching the door they were invited into a warm, cozy atmosphere. There sat Ed and Ernie in the middle of a card game with Doc, eating home-made sauerkraut and drinking home brew. They had decided to wait out the windstorm before coming back down the hill. Although they appreciated the concern, the two men were a little regretful that their party had been brought to an end. Three months later when the logging season closed, Ed, Dorothy, and Ernie left Falk.

A CALL TO ARMS

Time had come to a virtual standstill for the town. The world had passed Falk by. It seemed impossible that the fallen community would ever again be concerned with vital matters. Yet an international crisis erupted in 1941 which did, to some degree, affect the town. Outside the sleepy Elk River Valley a Second World War raged across two oceans and four continents. Falk happened to lie directly beneath the lanes of travel for west coast aircraft, and because the government feared a surprise attack on the mainland by the Japanese, it established a tight link of civilian-manned airplane-spotting stations along the entire West Coast.

Red and Ruth were asked to establish one of these facilities at their home in Falk. They took classes in Eureka and learned to identify 50 different airplanes. A small observation deck was constructed in an open meadow by their house, and they were given a direct telephone connection to San Francisco for calling in the identities of any aircraft that passed overhead. It seemed ludicrous to have grown up in the peaceful Elk River Valley, without even the luxury of electricity, and now lie awake at night listening for the invading Japanese Air Force! Nonetheless, the Braghettas maintained this spotting station for two years.

In 1944 Ruth and Red left Falk and moved to Eureka. Access to the townsite was blocked and the whole area was declared off limits to sight-seers and souvenir hunters. The general store was boarded up. But many homes, as well as the post office, the dance hall, the hotel-cookhouse, the gas station, the mill, and the logging camps, were all just left, opened and abandoned. Yet, within this framework of deserted structures, two or three of the old-timers unofficially continued their hand-to-mouth existence.

The valley's early developments were slowly dissolving. The first demolition of property had occurred in the early 1940s when the State Fish and Game Department blasted the dam at Falk, and returned the Elk River to its natural flow. This was done because the dam's fish ladder had collapsed and large schools of salmon which were attempting to spawn were unable to pass the deteriorated structure.

In 1949 the Bucksport and Elk River Railway abandoned its tracks to Falk and then sold the entire company the following year. The railroad's new owner was the Pacific Lumber Company, which also purchased all of Dolbeer and Carson's timberland on the North Fork, and continued to log in that area. Two years later the new owners of the

Richard "Dick" Sirignoli stands on the front porch of his cabin in Falk shortly before it was destroyed by fire in 1961. Dick was the last person to live in Falk. (COURTESY OF FRED ELLIOTT)

railroad served notice of their intent to abandon the line. This forced the Elk River Mill and Lumber Company, now only a property holder, to dismantle their logging railroad and scrap their heavy equipment. At a future date, without the main valley railroad intact, the company wouldn't have been able to salvage anything.

A dismantling train entered Falk and its crews proceeded to tear up the logging railroad. They salvaged two logging tractors, several spool donkeys, a bull donkey, a pile driver, the two locomotives, and tons of winches and cables, plus the entire railroad from Falk to the North Fork junction. Early in 1953 the Pacific Lumber Company's log train made its last run down the valley, and by May of that year, the Bucksport and Elk River Railway was totally dismantled by contractor Fred Botsford.

With the death of the railroad the entire valley experienced a quiet that had not existed since 1886, when the first trains had run. During those past seven decades, a story had unfolded that began in a virgin redwood forest, with the dreams and ambitions of two men. The saga that had followed drew a cast of characters from across the country and around the world. Their paths of discovery converged in the Elk River Valley. They had struggled together through hardships, joys, love, and tragedy, and created a small community which flourished for three generations. But in the end, when these efforts had succumbed to economic and historical change, nature began to slowly recover the townsite that had been carved from its forest.

In 1953, following the death of the valley's railroad, there remained only one person to shuffle among the decaying buildings and keep nature at bay. Richard "Dick" Sirignoli had labored in Falk's woods since 1907. He was an Italian immigrant, a very hard-working man who believed in his traditions. He had come to this country to seek opportunity, and he had found that in Falk. Through the many years that he had worked there, Dick habitually sent all of his saved earnings back to Italy so that one day he could return to his homeland to spend his final days. The Depression had caught him unprepared, and when he could not find another job, Dick wired his bank in Italy to request some of his funds. But social conditions had taken some dramatic changes in his homeland. Mussolini had come to power. The bank's return telegram informed him that his life's savings were now state property. Shocked, but determined to survive, Dick Sirignoli prepared to make the best of his situation and live his life close to the land. For 24 years after the mill closed, Dick fished, gardened, hunted, made his own brew, and endured all difficulties that came his way, until 1961, when his cabin burned to the ground. Dick escaped the fire but he chose to move on rather than to rebuild. His departure removed the town's last human inhabitant. The forest had finally reclaimed the ghost town of Falk.

FRAMEWORKS

On the previous page, looking out from the train barn at Falk,
one of the few remaining structures of what was once a
*thriving mill town. (*PHOTO BY JON HUMBOLDT GATES*)*

"All of the pictures made my memories return. I began
thinking about my life in the valley. It's so real to me now. I
look back and it just seems like yesterday. We were poor, but
the things we could do that kids today can't do! We could walk
far into the woods and gather trilliums and blackberries, we
attended the little one-room schoolhouse and our own community
church. That's what formed the foundation of what I believe
in today—simple teachings. I think of all those people and how
those experiences have really stayed with me. We were so lucky
to have lived like that. I've been thinking, I suppose it's all
there, the way I remember, the river, the violets, the huckleberries,
and who knows, with re-forestation, maybe there will be
another village there someday. But it would probably be all
trailer houses. It'll never be like it was in 1907."

Lelah Wait
1982

Above, the forest slowly consumes the remains of the Elk River Mill. Left, the house where Jonas Falk lived, overlooking a once-busy mill town. (TOP PHOTO BY DENNIS SULLIVAN; LEFT BY SY BEATTIE)

Right, empty shelves and knocked-over cases in the abandoned general store. Below, a truck inside a garage, slowly rotting together. Opposite page, a pair of boots and a hook, waiting for the workman who will never come. (TOP PHOTO BY JON HUMBOLDT GATES; PHOTOS BELOW, OPPOSITE BY SY BEATTIE)

Epilogue

ONE HUNDRED YEARS HAVE PASSED SINCE NOAH FALK
came to the Elk River Valley and twenty-one years since Dick Sirignoli
left. I have been listening to the sounds that now occupy the town. There
is the occasional sound of birds and rustling tree branches, but most
dominant is the Elk River. These are the same sounds that Noah and C.
G. Stafford probably heard when they stood in the forest surveying this
building site a century ago. Where thick ferns and lush foliage had
grown, planked roads and wooden walkways were laid. Where a great
forest had stood, buildings and railroads were constructed. Where a river
had flowed freely, a dam was built. Now, nearly five decades after the mill
closed, the river is again running unobstructed, and small trees dominate
the area where industry had flourished. The wooden walkways have
returned to the soil, and in their place are narrow dirt paths which are
used mainly by wildlife.

Since my first visit here, 14 years ago, many changes have occurred. While walking up the valley, I once again approached the site of
the two homesteads located on the outskirts of Falk. All that remained of
Charlie and Loleta Webb's home was part of a fence, a half-burned wood-
en water tank, and a vacant foundation of bricks and pipes. The Webbs had
died several years previous, and their home had since been destroyed. Their
neighbor, Stedman Fryers, died in 1981, and his home was also gone.
The metal gate which had been the boundary of the townsite had been
moved and now included the two homesteads.

A chill passed over me while I stood on the small cement walkway
that once led to the Webb's front door. On numerous occasions they had

invited me into their home and we sat in their parlor having tea and biscuits while they told me stories about their past. But now they were gone, and their home no longer existed. Yet, as I thought about those visits and remembered the relaxed, friendly atmosphere in which our conversations took place, I realized that even though the Webbs had passed on, some of their life experiences had survived through their stories.

Farther along the path toward the town, more change was apparent. Several homes still clung to the steep hillside, but they were now being digested by the encroaching plant life and were barely visible. The town itself could no longer be seen. Its buildings had sunk to such a ruinous condition and contained so many hazards that in 1979, the owners had decided to destroy the town entirely. Piles of decaying lumber are its final testimony. In a meadow, across the Elk River, the locomotive barn still stood—the only company structure to withstand the elements thus far. During my walk through the townsite, I felt that this would possibly be my last visit to Falk, so I decided to spend the afternoon in the train barn. It was from within the walls of that century-old building that I wrote these words.

The life of the early-day timber workers was hard, and for them few benefits had existed. Retirement and social relief programs were not yet established, and, in general, society's institutions lacked compassion for their personal hardships. Some of the lumber barons enjoyed lives of luxury, but there were few comforts for the people who gave their muscle and blood to harvest the redwood forests.

Despite this, when talking about their lives and all that had come to pass since the turn of the century, the people of Falk remembered a different side of that era. They had been bound together by camaraderie and friendship, and had lived a life that afforded only the basic needs. They spoke mainly about human concerns; their families, their friends, their fellow workers, and the community they had shared.

The story of Falk and its people provides a useful lesson for today. People now have more security and material comfort, but they have become scattered and isolated, connected to their friends and families only by the telephone and automobile—a marked contrast to the close community of Falk. Reflecting on this, several of the old-time loggers that I talked to shared with me some of their impressions of what has evolved within the timber industry itself since the early logging days.

Years ago, manpower was the mainstay in the mills and forests of the redwood region. Crew members, besides laboring next to each other

12 hours a day, six days a week, often shared cabins and meals, and even got together during their off hours. The men worked as a team, sharing the loads, helping each other through difficulties, trading stories, insults and laughs, and sometimes being forced to bear tragedy when one of their close friends fell victim to the surrounding dangers. These early loggers lived as brothers in their work.

But some of them view the modern timber industry with mixed feelings. They still see it as a dangerous occupation, but they are aware that technology is slowly replacing the rank and file worker in the woods and mills. The invention of the chainsaw alone now allows a single woodsman to cut over 100,000 board feet of timber in a day. The early tree fallers, with their axes and handsaws, could cut only about 10,000 board feet a day between two men. One old timer expressed his concerns to me about the new technology as we talked in the living room of his Eureka home one day.

He had been watching modern logging and milling operations develop over the years and had begun to see more people working by themselves. In some cases workers were performing their jobs while totally enclosed in a machine—operating huge cats, loaders, yarders, conveyors, log derricks and various other mechanical implements of the industry. He had seen some of the milling procedures which were now controlled by one person, who, from a small viewing booth filled with buttons, lights and levers, could manipulate a log weighing tons with the flick of a wrist. The old logger agreed that sophisticated computers and machinery could increase production and efficiency, but he also perceived that these developments were replacing more than just jobs, they were replacing a lifestyle. He felt that a much more competitive feeling now existed between workers, in contrast to the team spirit of the old days. At one point in our conversation he leaned toward me and spoke softly:

> "You know, all this automation and modern invention is just getting people more isolated from each other all the time. These days it seems people are spending more time relating to machines and TVs than they are to each other. Hard to say, but things keep going like this, and in another 100 years, no one's going to feel like they have a place in the world anymore. Machines and computers will do it all."

While looking up at the high, open rafter ceiling of the locomotive barn, my memories of the old logger, and his far reaching observations, slowly began to dissolve. I could see particles of sunight and patches of blue sky showing through the building's rusty and disappearing corrugated metal roof. I realized that Falk, and all it stood for, would be preserved only in memories and photographs, for nature and civilization had conspired to wipe out the last vestiges of the town and the way of life it had represented.

Although Falk had been but a brief occurrence in the history of modern development, it was a microcosm of the forces that have guided evolution since time began. We can look back at our earth's history and see early geological formations, the beginnings of the great forests, the age of dinosaurs, the expanding and retreating glaciers, and the civilizations of human beings, past and present. All of these phenomena share one common link—their existence has been subject to the rhythm of nature, which evolves, thrives, and in time, grows extinct, through a perpetual cycle that advances throughout all of creation.

With this thought in mind, I picked myself up off the dusty floor of the train barn, and with a last look around at the emerging forest, left what had once been Falk's Claim.

Jon Humboldt Gates

Appendix

PEOPLE FROM THE FALK AREA

Aho, Alli (Price)
Aho, Elizabeth
Aho, Emil
Aho, Joe
Albee, Eva
Albee, Link
Anderson, Guy
Barnes, Charlie "Pete"
Barnes, Chester "Ted"
Barnes, Elisha I
Barnes, Elisha II
Barnes, Elizabeth I
Barnes, Elizabeth II
Barnes, Emma (McManus)
Barnes, Jesse
Barnes, John
Barnes, Loleta (Webb)
Barnes, Marshall
Barnes, Ruby (Olson)
Barrie, Frank
Beauchamp, Charlie
Biord, James
Biord, Maggie
Birdsall, John
Bishop, Bill
Bishop, Ed
Black, Bill
Blaikie, Oliver
Blanton, Lee
Bledsoe, Earl
Blow, Bertha
Bowman, Billy
Bradford, Sam
Braghetta, Ellen
Braghetta, James
Braghetta, William "Red"
Brazil, John
Callihan, Arthur
Callihan, Ed

Callihan, Helen (Fleming)
Callihan, Lottie
Callihan, Shirley
Callihan, Zella (Moore)
Camenisch, Andrew
Campbell, L.H.
Cando, Henry
Carr, Johnny
Carter, Ivora
Carter, Mary (Wrigley)
Carter, Matt
Carter, Norman
Cavanaugh, Charles
Cavanaugh, Herman
Christie, Alma (Bonstell)
Christie, Clara
Christie, Elenore (Englehart)
Christie, Frank
Christie, George
Christie, Gladys (Smith)
Christie, Glenna (Smith)
Christie, Herbert
Christie, Ivan
Christie, Jet
Christie, Les
Christie, Lyman
Christie, Ruth (Johnson)
Christie, Ulysses
Clay, Jim
Cleasby, Everett
Cleasby, Kenneth
Cleasby, Velma
Cleasby, Wesley
Cloney, Jim
Coates, J.D.
Connally, Jack
Connally, Myrtle
Copeland, Jim
Crowley, Cecelia

Crowley, Lawrence
Crowley, May
Crowley, Neil I
Crowley, Neil II
Daily, Carrie
Danielson, Dewey
Danielson, Ed
Danielson, Erma
DeBeni, Frank
DeBeni, Ida
DeBeni, Larry
DeBeni, Louie
Dennison, Miles
Dinsmore, Mr. & Mrs.
Duckett, A.
Elliot, Lon
Elliot, Mert
Elliot, Ray
Emenegger, Frank
Emenegger, Josephine
Erickson, Ed
Erickson, Dorothy
Fahrney, John
Falk, Amelia
Falk, Besse
Falk, Blanche
Falk, Charles
Falk, Charles C.
Falk, Curtis
Falk, Dorothy
Falk, Drury
Falk, Elijah
Falk, Eugene
Falk, Harry
Falk, Irving
Falk, Jonas
Falk, Louise
Falk, Muriel
Falk, Nancy

North, Olaf
Norton, Charles
Norton, Flopper
Norton, Tom
Noyes, Ben
Oberdorf, Peter
O'Brien, Blanche
Olson, Elmer
Olson, Lenwood
Olson, Lewis
Olson, Rosie
Olson, Vernon
Palmer, J.
Parr, Euphemia
Parr, Lucinda
Parr, Ralph
Pease, Ed
Perona, John
Peterson, Jack
Quinn, Ed
Randall, Jack
Randall, Lee
Randall, Minnie
Richmond, S.
Rittenhouse, Stacy
Roberts, Donald
Roberts, Fred
Rollins, Elmer
Rollins, Jeanie
Ross, Clarence
Ross, Etta
Ross, George
Ross, Gertrude
Ross, Grace
Ross, Jessie
Ross, Pike
Rouf, Charles
Rowe, Albert
Rowe, Frank
Rushing, Evan I
Rushing, Evan II
Rushing, Dorothy (Edson)
Rushing, Grace (Henkell)
Rushing, Hazel
Rushing, Lloyd
Rushing, Wallace
Sarlund, Ed
Sarlund, Ernie

Sarlund, George
Sarlund, Oscar
Sarlund, Sunni
Scribner, Benton
Scribner, Chloe (Christie)
Scribner, Fred
Scribner, Leila
Scribner, Louisa
Scribner, Lloyd
Scribner, Seth
Shaw, Agnes
Shaw, Cathryn (Gilmore)
Shaw, Dorothy
Shaw, Frank
Shaw, Grace (Wrigley)
Shaw, John
Shaw, Joseph
Shaw, Jake
Shaw, Louise
Shaw, Mary (Wrigley)
Shaw, Robert
Sirignoli, Richard "Dick"
Spence, Joe
Steenfott, Charles
Steenfott, Ethel
Steenfott, Jimmy
Steenfott, Nathalie
Steenfott, Nedra
Steenfott, Norton
Steenfott, Valois
Stenka, Bill
Stockhoff, Sherman
Stuart, Adam
Stuart, Alvin
Swanback, Ernie
Swanson, Ruth (MacMillan)
Thrapp, Hiram
Townsend, Jim
Van Horn, Charles
Van Horn, Harry
Van Horn, Luella
Van Warner, Doc
Vinum, Curtis
Vinum, Elmer
Vinum, Mr. & Mrs. Helgar
Vinum, Victor
Waldorff, Annie
Waldorff, George

Waldorff, Lulu
Waldorff, Mollie
Waldorff, Wessley
Waldorff, William
Wall, Cap
Ward, Walter
Weatherby, Clarence
Weatherby, Gladys
Weatherby, Grace
Weatherby, Harold
Weatherby, Jack
Weatherby, Letha (Butterfield)
Weatherby, Mary
Weatherby, Nellie
Weatherby, Russell
Weatherby, Ruth
Webb, Charlie
Weismiller, Loren
Weismiller, Jane
Williams, Clara
Williams, Eli
Wheelahan, Hap
White, E.J.
Wilson, Alex
Wilson, Lena
Wilson, Sigrid (Kallio)
Wrigley, Alice
Wrigley, Bernice (Barnes)
Wrigley, Dorothy
Wrigley, Earl
Wrigley, Ethel
Wrigley, Everett
Wrigley, George I
Weigley, George II
Wrigley, George E.
Wrigley, Harold
Wrigley, Henry
Wrigley, Irving
Wrigley, James
Wrigley, Julia
Wrigley, Leila
Wrigley, Mary I
Wrigley, Mary II
Wrigley, Merced
Wrigley, Percy
Wrigley, Ruth (Braghetta)
Wrigley, Ted
Wrigley, Winfield

JONES PRAIRIE ELEMENTARY SCHOOL TEACHERS

1887	Miss A. L. Doe	1905	H. H. Johnson
1887	E. Jameson	1905-1906	Nellie Vail
1887	Rae Felt	1906-1907	Fred McCann
1888	K. McGowan	1907-1908	Grace Strobel
1888	Miss Hattie Highfield	1908	Oliver Petty
1888	G. W. Rager	1909	H. A. Kendell
1888	Ellita Mott	1910-1912	Luella Van Horn
1889-1890	Kate Newman	1912	Myrtle Barnum
1890	John Coats	1912-1913	Mary Weatherby
1890	Mrs. Ada E. Coville	1913-1915	Maude Frost
1891	Carrie L. Dickson	1916-1917	Mollie Gatliff
1892	Mary A. Dungan	1918-1920	Ruth Swanson
1892-1894	Myron Young	1920-1923	Maude Frost Andrews
1894-1895	Lizzy Quick	1922-1924	Ruth Mac Millan
1896	Inez Elsmore	1925-1926	Elsie Peterson
1896-1897	Maggie Laughlin	1926-1927	Marion Stuart
1897-1898	Alice Thompson	1910-1955	Alice Wrigley
1898-1901	Lizzy Quick	1955-1959	Gail Buchanon
1901-1902	Catherine Zane	1960-1963	Fred McGill
1902	Kate Stearns	1963-1964	Mary Lengel
1903-1905	Alma Christie		

Former residents of Falk gather at a reunion organized by author Jon Humboldt Gates, seated on the floor. Top row, Nathalie Baily, Norton Steenfott, Neil Price, Wayne Miller, Karla Wrigley, Irving Wrigley, Joe Aho, Helen Miller, Merced Wrigley, Ruben Forbes, Ruth Johnson, Shirley Callihan, Vernon Olson, Evan Rushing, Ted Barnes, Mary Barnes, Mary Wrigley. Seated, from left, Mrs. Ruben Forbes, Dorothy Erickson, Alli Price, Letha Butterfield, Helen Newell, Lelah Waite, Ruth Braghetta and Leland Newell. PHOTO BY SAYWARD AYRE

Bibliography

PERSONAL INTERVIEWS

BARNES, Ted and Mary. Personal interview with author, 1980, 1981.

BRAGHETTA, Ruth. Personal interview with author, 1980, 1981.

ERICKSON, Dorothy. Personal interview with author, 1981.

FALK Sr., Judge Harry. Personal interview with author, 1970.

FALOR, Frank. Personal interview with author, 1970.

FALOR, Gerald. Personal interview with author, 1982.

FORBES, Ruben. Personal interview with author, 1980.

FROST, Ralph. Personal interview with author, 1970.

JOHNSON, Ruth. Personal interview with author, 1980, 1981.

MILLER, Wayne. Personal interview with author, 1981.

NEWELL, Leland and Helen. Personal interview with author, 1980.

NISKEY, Charlotte. Personal interview with author, 1981.

OBERDORF, Peter. Personal interview with author, 1980.

OLSON, Vernon. Personal interview with author, 1980, 1981.

RUSHING Jr., Evan. Personal interview with author, 1980, 1981, 1982.

STEENFOTT, Norton. Personal interview with author, 1980, 1981.

SWANBACK, Ernie. Personal interview with author, 1982.

WAIT, Lelah. Personal interview with author, 1981, 1982.

WEATHERBY, Russell. Personal interview with author, 1981.

WEBB, Charlie and Loleta. Personal interview with author, 1968, 1969, 1970.

WRIGLEY, Mary. Personal interview with author, 1969, 1970.

WRIGLEY, Merced and Karla. Personal interview with author, 1981.

WRIGLEY, Irving. Personal interview with author, 1969, 1970, 1980, 1981, 1982.

WRIGLEY, Ted. Personal interview with author, 1980, 1981.

TED & MARY BARNES

RUBEN FORBES

VERNON OLSON

RUTH JOHNSON

EVAN RUSHING

LELAND & HELEN NEWELL

TED, RUTH & IRVING
WRIGLEY

RUTH BRAGHETTA

PETER & EVELYN OBERDORF

LELAH WAITE

MERCED & KARLA WRIGLEY

RUSSELL & MARY WEATHERBY

CHARLIE & LOLETA WEBB

NORTON STEENFOTT

WAYNE & HELEN MILLER

PHOTOS BY BEVERLY HANLY, TERESA HENDERSON AND THE AUTHOR

Books and References

Barraclough, G., Editor. "History of the Earth's Ages." *Times Atlas of World History.* Maplewood, New Jersey: Hammou, Inc., Publishers.

Carranco, Lynwood, et al. *Logging the Redwoods.* Caldwell, Idaho: Caxton Publishers, 1975.

Coy, Owen C. *The Humboldt Bay Region: 1850-1875.* Eureka, California: Humboldt County Historical Society, 1982.

Fagan, Brian. *People of the Earth,* 2nd Edition. Boston-Toronto: Little, Brown and Company, 1977.

Guinn, J. M. *History of the State of California and Biographical Record of Coast Counties.* Chicago: Chapman Publishing, 1904.

Heizer, Robert F., Editor. "Development of Regional Pre-Historic Cultures." *Handbook of North American Indians.* Washington, D.C.: Smithsonian Institute, 1978.

Irvine, Leigh H. *Humboldt County, California.* Los Angeles, 1915.

Melendy. *One Hundred Years of the Redwood Lumber Industry: 1850-1950.* Humboldt County Library, Eureka, California. Stanford University, 1952.

Onstine, Frank. *The Great Lumber Strike of 1935.* Topeka, Kansas: Standard-Hart Printing Company, 1980.

Palais. *Pioneer Redwood Logging in Humboldt County.* Humboldt County Library, Eureka, California. Compiled 1974.

Paul, Rodman W. *California Gold.* Cambridge: Harvard University Press, 1947.

Raphael, Ray. *An Everyday History of Somewhere.* New York: Alfred A. Knopf, Inc., 1974.

Smale, Rudolph. *There Go The Ships.* Caldwell, Idaho: Caxton Printers, 1940.

Spinney, Les. *Newburg.* Eureka, California: Schooner Features.

Other Sources

"Bucksport and Elk River Railroad Company: General History." Courtesy of Riggs Johnston. See also, Pamphlet Section, Humboldt State University Library, Humboldt Room, Arcata, California.

"Falk Family Genealogical History." Courtesy of Mrs. Glenn Schaffer.

"Falk." Pamphlet Section. Humboldt State University Library, Humboldt Room, Arcata, California.

"First Congregational Church of Elk River." Records. Courtesy of Ruth Braghetta and Eureka Congregational Church.

Fountain, Susie Baker. Notebooks. Susie Baker Fountain Papers, Volumes 11, 31, 47, 50, 62, 70, 82, and 85. Humboldt State University Library, Humboldt Room, Arcata, California.

Frost, Ralph. Personal Diary. Courtesy of Dorothy Bailey.

"Jones Prairie School Journal, 1898-1947." Courtesy of Mr. and Mrs. Ralph Krause.

"Jones Prairie Elementary School Records." Courtesy of South Bay Elementary School.

"Land Records, 1875-1890." Recorder's Office, Humboldt County, California.

"Nellist Stage Lines Log, 1914-1918." Courtesy of Norton Steenfott.

O'Brien, David. *California Employer-Employee Benefits Handbook*, 6th Edition. Courtesy of Franklin Grady. Department of Industrial Relations, State of California.

"Production and Shipping Records: Elk River Mill and Timber Company, 1910-1920." Courtesy of Richard Billington.

"Requisitions: 1882-1900." Registers of School Superintendent, Humboldt County Library.

Roberts, Earl. "A Story of Industry." Collection of local historical records involving the timber industry. Courtesy of Daisy Roberts.

"Study of Cultural Resources in Redwood National Park." Polly Bickel, 1979. Redwood National Park Office, Arcata, California.

Willis, R. G. Notes. "Labor Movement in Humboldt County: 1893-1910." Humboldt State University Library.

ARTICLES AND PERIODICALS

Among the newspapers and periodicals consulted in the course of the author's research were: *Humboldt Historian*, the publication of the Humboldt Historical Society; *Science* magazine; *The Timberman*, a lumber journal published in Portland, Oregon; *Western Railroader* magazine; *Zoological Society Bulletin*; *Humboldt Times*, weekly; *Humboldt Times Newspaper*; *Times-Standard*; *Humboldt-Standard*.

SPECIAL THANKS TO THOSE WHO SUPPLIED ADDITIONAL INFORMATION, PHOTOS AND ASSISTANCE

Joe Aho
Paul A. Asp
James H. Baker
Mr. & Mrs. Joe Barlow
Sy Beattie
Jim Bernard
Mrs. Evadne Bradley
Letha Butterfield
Shirley Callihan
Wilma Calvert,
  Fortuna Historical Museum
Cheney Lumber Company
Carl Christensen
Chalmers Crichton,
  Humboldt Historical Society
Robert Crichton
Coleen Kelley,
  Clarke Memorial Museum
Dorothy Edson
Fred Elliott

Red Emmerson
Bill Fahey, Fort Humboldt
Harry Falk III
Ricks Falk
Keith Fryers
Alan Grau
Steve Hart
Benjamin F. Harville
Jack Hitt
Helyn Johnston
Gerald Kaminski
Joey Mallett
Jack McManus
Evelyn Oberdorf
Peter Palmquist
Stan Parker,
  Pacific Lumber Company
Phillips Camera Shop
Neil and Alli Price
Oregon Historical Society

Martha Roscoe
Erich Schimps, Humboldt
  State University Library
Gino and Helen Scuri
Ann Smith,
  Redwood National Park
Henry Sorensen
Les Spinney
Laurey Sullivan
Sam Swanlund
Jim Test
Times Printing Co.
Gloria & Paul Wills
Mark Wilson,
  Humboldt State University,
  Forestry Department
Rod Wooley,
  Pacific Lumber Company
Dorothy Wrigley
Jim and Veda Wrigley